Suamico, The Fishing Village

Boyhood Memories Of Suamico, Wisconsin's
Commercial Fishing Past
And The People Who Lived It
(Circa Mid 1950's)

By

Dan Williquette

Paisa (Alt) Publishing Co.
De Pere, Wisconsin

Copies of “Suamico, The Fishing Village” are available from:

Dan Williquette
730 Pinecrest Rd.
Green Bay WI. 54313
920.434.4866

Published by Paisa (Alt) Publishing Co.
502 George St.
De Pere, WI. 54115
920.983.5326

Printed in the United States of America
Library of Congress Control No.2007938525
ISBN 0976478285

Cover by Dan Williquette
Front cover art by Artist Agnes Wainwright

Contents

Dedicated to Dick Devroy
For a lifetime of fun.

Acknowledgements

To those who contributed to the contents of this book;

My Wife Sandra
Dick and Teresa Devroy
Bill Devroy
Jim Schiltz
Ron Schiltz
Darold Rogers
Bobby Komorowski
Marilyn Posluszny
Dick and Cathy Gokey
Lyle Kirby
Val Drzewiecki
Eddie and Barb (Kozloski) White
Jan Noe
Allan and Dianne Dombrowski
Joe Jr. (Scooter) and Marion DeWitt
Francis Euclide Family
Tom Valentine
Bill Dickinson
Dan Laubenstein
Bob Nelson
George and Angie Tuttle
Jim Cleveland
Richard Cleveland
Eileen (Posluszny)Terrien
Joan Mutz
Monica Mutz
Donald Valentine
Tom and Norma Peters
Mark and Nick Maricque
Barb Drzewiecki
Donnie Benedyk
Lee and Glory Hermes
Nancy (Brunette) Krause
Everett Marks
Eugene Marks
Mike and Kathy Williquette
Marie Zakowski
Julia Posluszny
Betty Dombrowski
Curt Nelson
Marcie Ertel
Eileen (Malchow) Olson
Geri (Kozloski) Van Lanen
Agnes Wainwright
Norbert and Marion (Borowitz)Vandersteen
Chris and Betty Lou (Valentine/Vervoren) Conrad

The Cover

The cover is a painting I found at a flea market at St. Norbert College. When I saw it I couldn't believe my eyes. It's a painting of the mouth of the Suamico River looking east from where the public boat landing is now. The boat in the foreground is the "Arbutus." Dick Devroy owned it at the time. He remembers the woman painting it. There were several ladies painting at the same time, he thought they were schoolteachers.

As it turns out the lady's name is Agnes Wainwright. Norma and Tom Peters own a painting by her also. Norma has researched her background and learned she was a very prolific painter in the 1930s, 40s and 50s. She graduated from Green Bay East High School and attended the Chicago Institute Of Art. In 1945 she was a curator at Neville Public Museum. She has since passed away.

On this particular day Dick noticed that one lady wasn't painting. He and his brother Bill had just gotten in off the Bay and for some reason had parked the boat the opposite way it's depicted in the painting. He asked why she wasn't painting and she said they had parked the boat the wrong way. He obliged her and turned the boat around.

The boat directly behind the Arbutus was owned by a Dr. Babcock from Pulaski. I've been told it was primarily used for duck hunting and burned up on the Bay. The third belonged to Melvin Wright. It was a drop net boat he purchased from Paul Lade. He added the cabin and converted it for pleasure use. The boat furthest back is the "Judy," Herb and Arnie Zakowski's drop net boat. I know it's theirs because it was the only boat on the river with a red stripe down the bow.

This is a picture of the Suamico River taken from the public boat landing. It's roughly the same spot where the cover picture was painted.

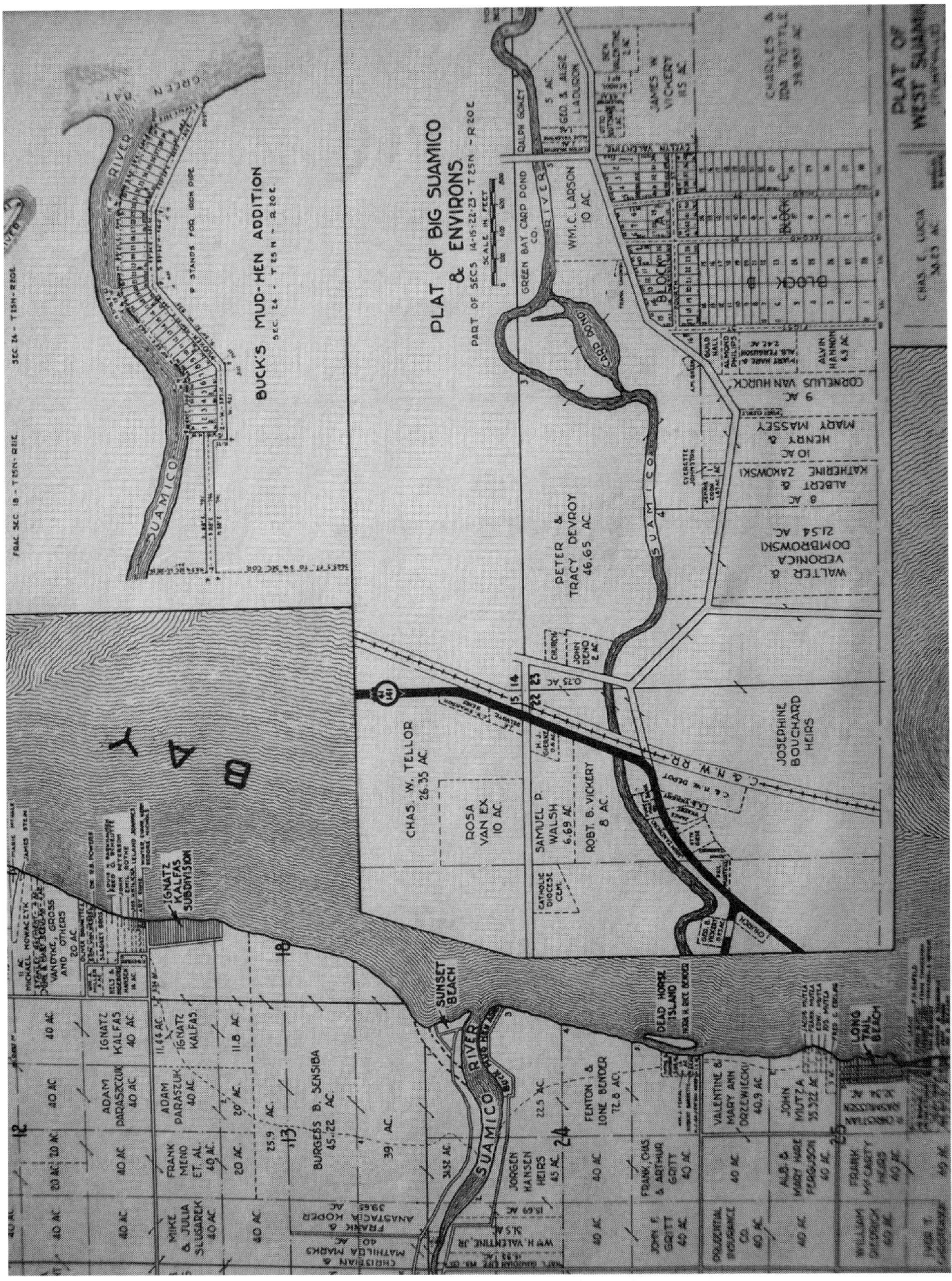

1936 Plat Map of Suamico, Wisconsin

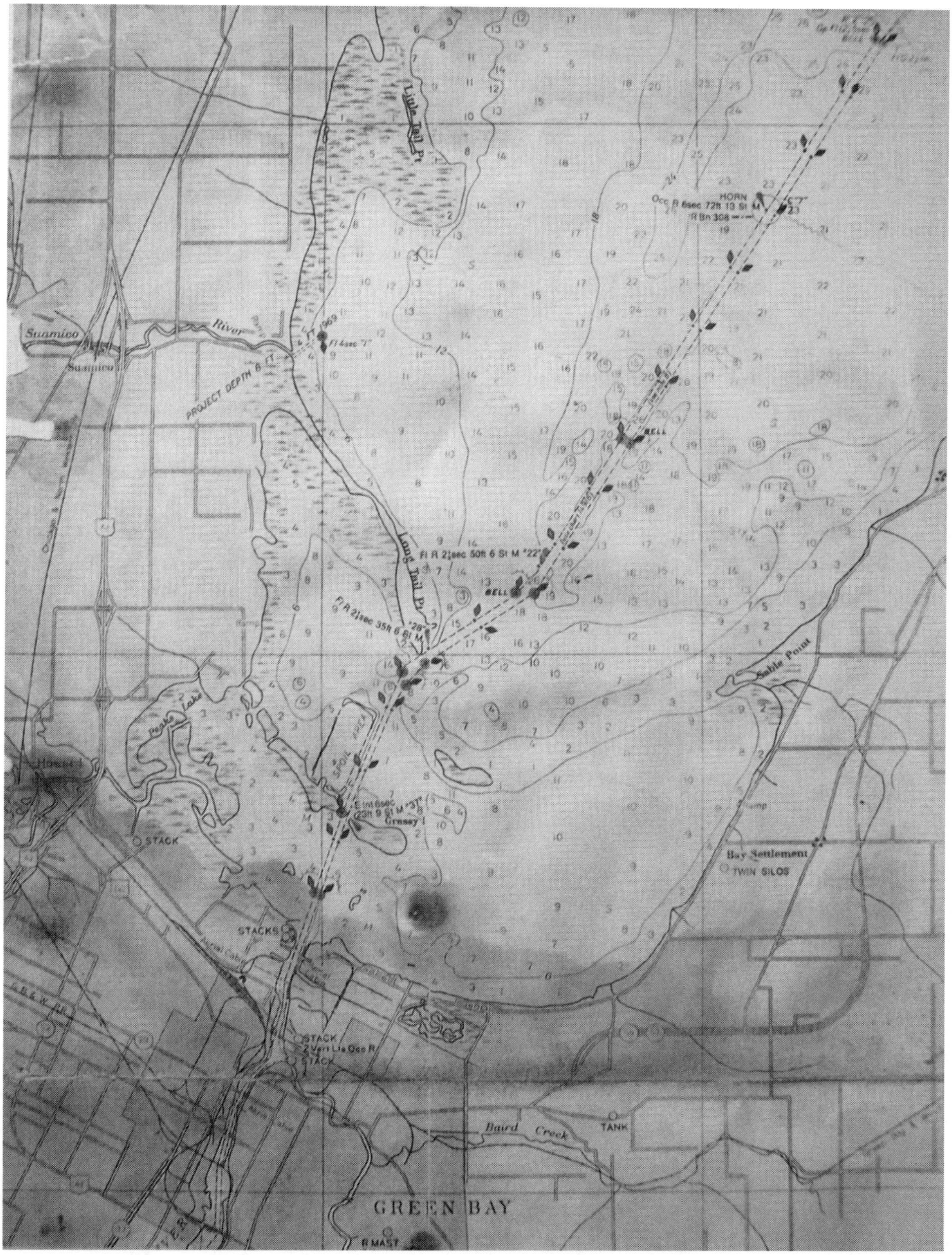

A navigational map of the lower end of Green Bay where most of the Commercial Fishing out of Suamico took place.

Suamico, The Fishing Village

The Village of Suamico wasn't always just a name people passing through couldn't pronounce and fields of $300,000 homes. When I was a kid, back in the 1950s, all the houses from the Methodist Church to the mouth of the Suamico River wouldn't add up to $300,000! But that area was what Suamico was all about; Commercial Fishing. Oh there were farmers and mink ranches, a box factory and a few people who worked in Green Bay, but commercial fishing was the main means of making a living.

Starting from a block south of the Methodist Church, (the intersection of Lakeview Dr. and Riverside Dr.), there were the Wright brothers, Harvey and Martin. Across the street from them was Francis Euclide.

"Down the road", (anything east of the Church), was Lyle Kirby, Art Devroy and his sons Bill and Dick; Normal Williquette and sons Dan, Mike and Jim; George Lince, John Valentine, George Ladrow, Bill Dickinson, Reggie Netoles, Ben Valentine, Jack Mealy, Bill Drzewiecki and his son Val, Ray Tuttle and his son Donald, Joe and Louis DeWitt and Joe Jr.; Johnny and Ralph Bender, Russell Devroy and his sons Ronnie and Bobby. On the north side of the river was Otto Marks and his sons Eugene and Everett, and Harold Devroy.

"Up the road", (anything west of the Church), was Dempsy Brunette, Leonard and Charles Dombrowski, Herb and Arnold Zakowski, Anton Borowitz and his sons Joe, Vic and Ted, and George Tuttle Sr. and George Jr.

Oliver Brunette and his sons Nubbie, Micky, Melvin and Walter lived north of the river between Listle's Resort and Jenk's Channel Resort at the end of what is now called Resort Road. Neither place is a resort any more. The Jenks property is now a sailing club. Jake and Jack Mutz were south of the river, north of Long Tail Beach.

These are the people I recall were fishing during the time frame this story is being written, circa mid 1950s. There were many before and after and I apologize if I missed anyone. Many of the names mentioned above can be found on the 1936 plat map of Suamico.

I didn't start out with the intention of writing a book. But then I thought, I've got all this material and all the memories, I might as well get it on paper before people forget the important role commercial fishing played in Suamico's past. But first, I'd like to get you acquainted with the Suamico I grew up in.

For those who might not know, Suamico is pronounced "swa-me-co", sometimes mispronounced "soo-a-me-co" by passers-by. It used to be called Big Suamico, because ten miles to the north is the town of Little Suamico, no relation. It's had many translations from "Tail of the Beaver" to "Big Sandbar" for Long Tail Point where Big Suamico is located and "Little Sandbar" for Little Tail Point where Little Suamico is located. I don't know if anybody knows for sure. It's probably of Indian or French origin or a combination of the two. But we've come to settle on just plain Suamico.

Suamico is about 10 miles north of the city of Green Bay. It used to have a sign, "Gateway To The North", and "Burma Shave" signs along the highway. Also a sign, "115 Miles to Mertins Café", it was located in the U.P. (Upper Peninsula of Michigan). They had mileage signs all over the country. They were rumored to have one in Russia.

There was a commemorative sign honoring all of Suamico's war heroes. It was located "up the road" next to the highway. I had forgotten about it until Lyle Kirby reminded me of it over a cup of coffee one morning. He used to visit it often because his Father's name was on it. He died in 1945 during the Battle of the Bulge in WWII. His name was Lyle also. I don't know what became of it, we should erect another one.

At one time there was a cardboard sign placed under the railroad viaduct that read, "38th Parallel." It was put there as a joke by warring school board members to show the dividing line between those who wanted Suamico kids to be bussed to Pulaski High School and those who preferred bussing them to Green Bay. They used the term "38th Parallel" because the controversy was taking place during the Korean War. They compromised and left it up to each family to decide where they wanted to send their kids. People used their heads back then instead of the courts.

Downtown Suamico was like most rural communities, with a few quirks. It had the usual complement of three churches and three taverns; plus it had three gas stations, two general stores, a railroad station, a post office, a three room school house, a guild hall, a ball field, a couple of restaurants, three mink ranches, a box factory, a bowling pin factory and the quirks; three fish houses, a boat builder, four carp ponds and four bridges that crossed the Suamico River within a mile of each other.

Dairy farms surrounded the town. Three resorts lined the shoreline. The old timers would tell of Dillinger driving through on his way to his hideaway up north. Velp Avenue was Highway 41/141. County J was the "lower road." You could find arrowheads just about anywhere. There were signs "Tractors With Lugs Prohibited" put up after they paved the roads. A paper mill byproduct of the pulp making process called, "spent sulfite liquor", was spread on the gravel roads under the guise, "road binder." It served two purposes; one, it kept the dust down, and two, it was a way for the mills to get rid of it. It also served a third purpose; it contributed greatly to the pollution of the Bay, because sooner or later, that's where it would end up.

Before all the pollution, the Bay flies would be hip deep under every beer sign in the morning. There were so many frogs, that at certain times of the year the roads would get so slippery, from running them over, you had to slow your vehicle to a crawl, especially on County J past the Barkhausen Game Preserve. We didn't have the many songbirds that are so common today. Thankfully, we've come a long way towards eliminating much of the air and water pollution that was overtaking the environment then. Unfortunately what we failed to do though, is halt the invasion of exotic plant life, especially the Phragmites that has overtaken our shorelines, and the sea creatures that have invaded our waters. Where cattails and lily pads once thrived we now have bamboo and pampas grass flourishing. Through our carelessness the whole eco system has had a change of face. It's unrecognizable from the way it looked a half century ago.

And we didn't have a lot of the things we take for granted now days. There was no rescue squad. When my Father died of a heart attack, there was nothing anyone could do. Polio was another thing. We were pre-vaccine. Everett Marks has suffered a lifetime with its crippling effects.

We grew up in the heat of the cold war. There were designated places to take shelter in case of an air raid. Larry Moore went so far as to build his own bomb shelter. At the Reforestation Camp, now the N.E.W. Zoo, we used to man the observation tower. It was totally enclosed back then with glass windows all around and had a direct line to the Civil Defense Headquarters in Green Bay. We were to report any plane we'd spot; number of engines and which direction it was flying. Our identification code when we'd report in was, BROVACO BECK 53 BLACK. It's funny how things like that stick in your memory. It was a volunteer effort through the Boy Scouts and other civic organizations and on an individual basis.

In the summer we had picnics. The Catholic Church had one every year, and still does.

The VFW used to have one at Vickery's Park, it no longer exists, and then at their clubhouse, it's gone too. My bike got ran over at one of the picnics.

Any picnic worth its salt (and pepper) features "Chicken Booyah" on the menu. For those unfamiliar with that term, booyah is basically chicken vegetable soup. Without it, a picnic is just a picnic. My grandpa, Ed Williquette was the booyah guru back then. He had a bag of secret spices he'd hang in the kettles while it was cooking. His great grandson, Jimmy Nier, has become his successor. It's pretty much a Green Bay area thing; nowhere else will you find people willing to drive for miles to sit at a picnic table in the hot summer sun and eat a bowl of scalding hot soup while wiping the sweat off their foreheads. It's probably why the beer tent is such a popular place.

Suamico and Duck Creek had a very spirited baseball rivalry. My Dad managed the Suamico team and his brother, my Uncle Bert managed the Duck Creek team. Many of the guys who fished commercially played on the Suamico team: Donald Tuttle, Horace Drzewiecki, Scooter DeWitt, Chuck Dombrowski, Billy Roderick to name a few.

They're all smiling so it's hard to tell which team had to buy the beer.

This picture was taken around 1950. In the forefront on left, are Norm (L) and Bert Williquette. In back row (L to R), unknown, Chuck Dombrowski, unknown, unknown, Jim Roderick, unknown, Billy Roderick and daughter Cathy, John Valentine and daughter Sandra. The picture was taken outside Blake's Bar in Duck Creek. In case you don't know where Duck Creek (Howard) is, it's located between Green Bay and Suamico.

I didn't even know there was a post office, until somebody told me. We were R.F.D. and never had to deal with it. I'm told Ambrose Sheedy was the postmaster and that he was a very cheerful guy who liked to whistle a lot. Our mailman was William (Doney) Valentine; he was all right too.

Jim Vickery ran the general store. You could buy groceries for yourself and a bale of hay for your horse. It was a Fleet Farm before its time. Jim was always there if you were a youngster looking for work. He always paid in silver dollars and always called the men "Son", no matter how old they were. His wife Edith had a flower shop next door. Every Easter and Mother's Day we'd get mom's corsage from her.

The train depot was a quaint place. Otto Burkhardt was the stationmaster. He always smoked a corncob pipe. We'd visit him before our Boy Scout Meeting, which were held across the street. It was an old wooden building with creaky floors, a potbelly stove and a few wooden benches. When everything was quiet, the only sounds you'd hear were the old Railroad Regulator Clock ticking and Morse code messages coming over the telegraph. It was so relaxing; I don't know how he stayed awake.

Living "down the road", we did most of our grocery shopping at Nina Green's Grocery store, later Martell's. I was always running there for a nickel's worth of yeast for Ma to bake bread. They carried some hardware item, nails, nuts and bolts, but for bigger items we'd have to go "up the road" to Vickery's. Friday nights were for shopping in Green Bay; that's the only night stores stayed open till nine. You could shop till noon on Saturday, but stores were closed on Sunday. We'd get our meat at Calewarts' Meat Market and freeze it in a rented locker at Otto Lehner's. The milkman would deliver our milk and we'd get chickens from Louis Slusarek and butcher them in the back yard. They'd run around like.........

We had a very robust Scouting program. We were Troop 33. Bill Vickery was the Scout Master and Pat Chambers assistant Scout Master. We played capture the flag in McNeill's sand hills and went to Bear Paw Camp in summer and winter. We even had an indoor rifle range below Kapla's garage.

Vince Schiltz was the Cub Master. Some of the Den Mothers were Thea Moore, Mabel Campbell, Edna Malchow, Lillian Stutleen and my mother Madeline Williquette. The Stutleens volunteered a piece of their property at Crooked Lake for a Cub Scout Camp; we named it "Camp Mohawk." I can never forget a skit one of the dens put on at a campfire. A kid came running through hollering, "the viper is coming, the viper is coming"; soon another kid, and then several more, all with the same message. Finally a kid with a roll of toilet paper came through hollering "I de viper, I de viper."

This picture was taken at a Blue and Gold Banquet at the Guild Hall. Front row, (L to R) Dan Williquette, Darold Rogers, Vince Schiltz, Ronnie Schiltz, Ben Stutleen. Back row Lillian Stutleen, Greg Chase, Jim Cleveland, Dick Leanna, David Valentine and Mabel Campbell.

In 1951 the Explorer Troop set a national record for promoting the most scouts to the rank of Eagle Scout at one time. 16 Scouts plus the Scoutmaster received the award at a Court of Honor held at the Guild Hall. In 1971 the troop broke its own record when it promoted 28 Scouts to the rank of Eagle.

The 1951 Court of Honor.

Front Row (L to R); Scoutmaster Bill Vickery, Richard Johnson, Robert McNeill, Larry Valentine, Donald Cann, Robert Wincensten, Alan Dombrowski.

Mid. Row; James Peters, Robert Dickinson, Vernon Malchow, Glenn Valentine, Dale Leanna.

Back Row; Robert Elliot, Allen Gokey, Ora Cahoon, Robert Stutleen, Valentine Drzewiecki.

So over-all Suamico wasn't much different than any other small town, rural community. What made it unique to this area, and probably the entire Lake Michigan shoreline, was the number of its residents, per capita (Suamico's population was roughly 1800 at the time), who relied solely on commercial fishing for their livelihood. From a high of over 30 commercial fishing boats lining the Suamico River at one time, to a much lesser number today, it truly deserves to be remembered as, "Suamico, The Fishing Village."

Making Model Boats

When I retired in 1997, I wanted to carve a model of the wooden fishing boats from that era to display in my rec. room. The wooden boats eventually gave way to steel boats. As the wooden boats became obsolete they were either burned or just allowed to rot away. I found the last surviving wooden boat in Vickery Village, a retail center made up of old historical buildings from Suamico. Val Drzewiecki donated the boat to become a museum piece, but that never happened and now it's just rotting away like the rest. Luckily it lasted long enough to get the measurements I needed to carve the model. The boats had beautiful lines and were unique to this area. Nowhere else do you find boats configured this way and built for the express purpose of fishing drop nets.

Drzewiecki's boat, the "Suamico."

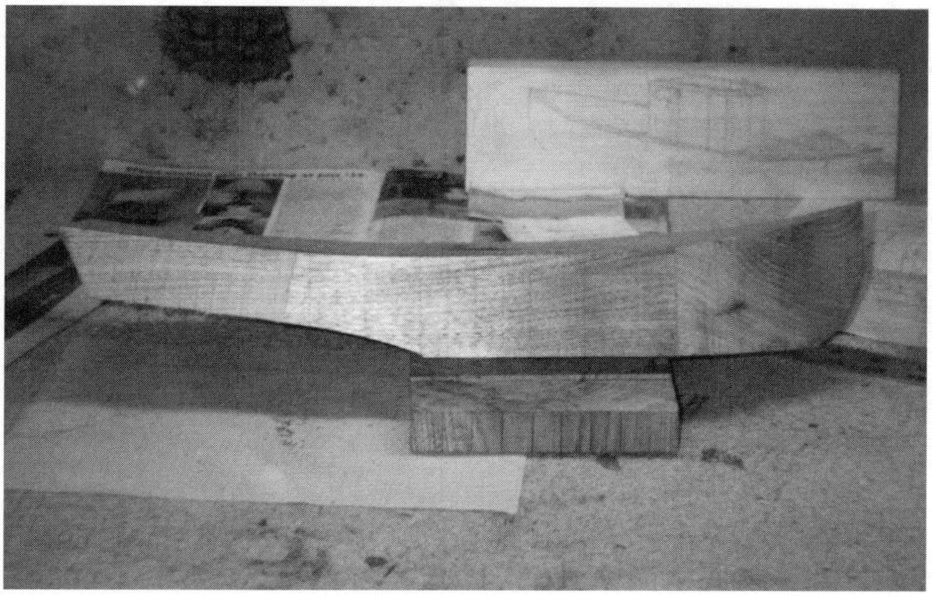

The start of a model.

After carving the first one I got hooked. I had to make one for Dick Devroy. My father died when I was 14. He stepped up and gave me and my two brothers, Mike who was 9 and Jim who was 3, the male influence we needed growing up. Sometimes good influence, sometimes not so good, but we didn't turn out too bad.

I thought one would look good in Val Drzewiecki's Suamico Fish Co., so I traded the first one I made to him for some fish.

The third one went to Bill Dickinson. He and I were partners at one time. We fished for perch, smelt and seined for carp. Dick Devroy gave us the name, “The Wallygoo Fish Company.” He was Wallygoo 1, and I was Wallygoo 2.

The fourth one went to Johnny Bender. He was dying of cancer and I wanted to do something nice for him. As sick as he was he still had his wonderful sense of humor. I had carved an anchor out of wood for the boat and he said, “It won’t work!” I thought I had carved it correctly so I asked him why it wouldn’t work? He said “cause it’s made of wood and would float!” His boat is the “Andy.”

The fifth one went to Francis Euclide. He was in a nursing home and as deserving an old fisherman as there was. It was a joy to visit him and listen to his stories. He and my Dad fished together for a while. He built a huge garage and the upstairs was devoted to making nets. It was a great place to go because sooner or later any old fisherman in town with nothing better to do would show up and the B.S. would start. I can still smell the wood fire and the new twine soaked in e-oil, a preservative.

The sixth one is a gill net boat. It is patterned after Paul Spott's boat the "Betty Anne." Paul got cheated out of a lot of life. He died way too young. He and his wife Betty were partners in life and in fishing.

Lyle Kirby got the seventh one. He keeps the fishing lore alive. It's his favorite subject. Today he raises perch for a living. I guess he got tired of going on the Bay to catch them. Now he just has to go to his ponds. On second thought, I'm sure if Lyle could relive those days on the Bay, he'd do it in a heartbeat.

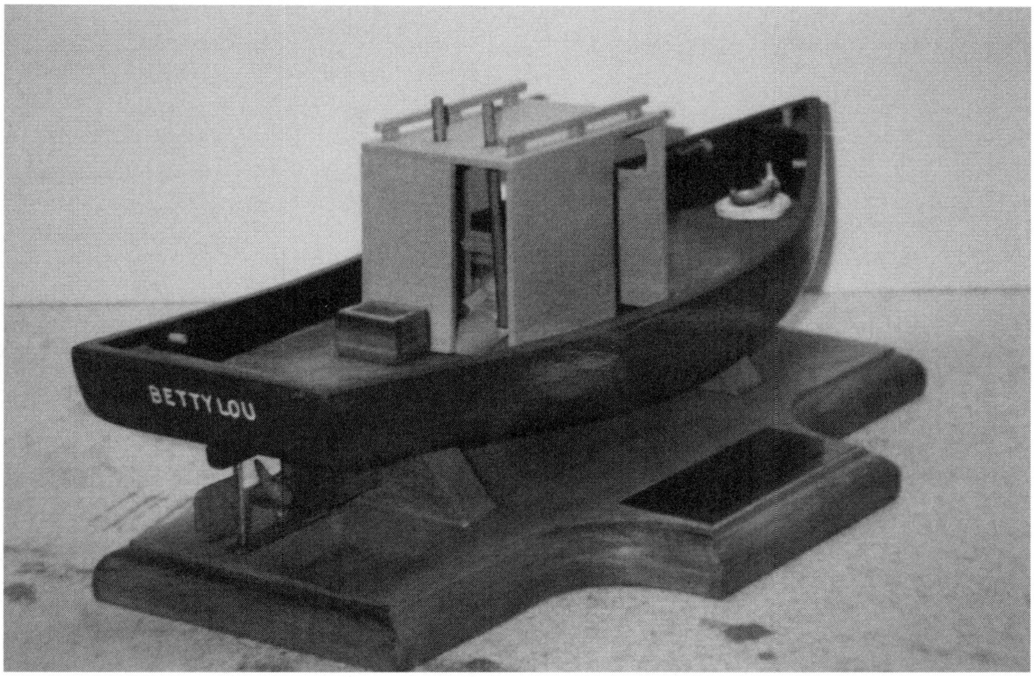

The eigth one I kept and named it the "Betty Lou." It belonged to my Uncle John Valentine. He and my Dad were partners for many years. John was married to my Dad's sister, Mercedes.

The latest one I made is promised to Marilyn Posluszny. She helped me add to my picture collection of Suamico fishing boats. She had pictures of Zakowski's boat the "Tateter." She also had pictures of the "Amelia," originally Drzewiecki's gill net boat, and the "Roy," Francis Euclide's boat being launched, probably for the first time. She is also a world famous bird carver. We saw some of her work when we visited her in California. We made a deal, I would make her a boat, and she would carve a seagull for me. Her uncles are fishermen Herb and Arnie Zakowski.

Part of my collection of Suamico fishing boat pictures.

Early on my goal was to collect a picture of every fishing boat in the river when I was a kid. I came very close missing only 3 of 15 at the time. But my search produced enough other pictures for the nucleus of this book.

Drop Net Boats

The primary boat used for perch fishing in that era was called the "Drop Net Boat." They were made of wood and built by one of two local craftsmen, Herman and Paul Lade (lah-dee) of Little Suamico or George Lince of Big Suamico. The boats were basically the same design but had subtle differences. The keels were the big difference and one sat a little lower in the water than the other. To anyone who didn't know the difference, they all looked alike.

They were made with different types of woods; white oak was preferred for the infra-structure and redwood for the planking. White oak was readily available and could be harvested locally. The boats varied in lengths and widths but averaged about 30 feet long and 10 feet wide.

There were only two color combinations. I don't know if it was an unwritten rule or what, but they were either painted green with a white cabin or gray with a white cabin. How they arrived at those combinations I have no idea. The inside of the hull was treated with cre-osote to keep the wood from rotting and the outside was clad in tin. Usually before the green paint went on, a coat of "red lead" was applied below the water line to keep the tin from rusting. Red lead would compare to Rustoleum today.

A flat head Ford V-8 with a marine reduction gear transmission was very popular.

Engines ran the gamut from pure marine set ups like 6-cylinder Chryslers with reduction gear transmissions to regular car engines such as straight eight Buicks or Ford V-8s with a stick shift transmission and a clutch with third and reverse gear. Diesel engines came later with the steel boats.

Francis Euclide made the steering wheels. He was a lathe operator and did beautiful work. In some of the boats the steering operated backwards. To turn left, you turned the wheel to the right and visa versa.

The sides were covered with tin to protect the wood from the beating they'd get when lifting the nets. The tin also provided protection to the bow when breaking ice in the late fall. The river would freeze before the bay and they could fish until just before the bay froze over.

Every fall the boats would be pulled out of the river for the winter. Over time boards would rot and need to be replaced. Leaks would develop and new caulking needed to be driven between the planking. New tin would be applied where needed. All this preparation would take place before launching each spring.

The "Alan," owned by Leonard Dombrowski, the "Mary Lou," George Ladrow and the "Betty Lou," John Valentine, froze in overnight ice.

Full Cabin Boats

Before the drop net boat became the boat of choice, fishermen used what I've heard referred to as full cabin boats or open bow gill net boats. It was a flat bottom or tunnel bottom boat made for fishing various type nets in the shallow waters of Little Tail and Long Tail sloughs.

Tunnel bottom boats had a tunnel in the bottom that ran from the engine to the back of the boat. The drive shaft and propeller were tucked up inside the tunnel so the prop was just under the surface of the water to allow the boat to operate in very shallow water.

Typical open bow gill net boat.

When the herring fishing dwindled, the focus turned to perch. The perch were fished with drop nets and in the deeper water of the Bay proper. The full cabin boats weren't designed to lift drop nets so they were used to tow a rowboat to work out of until the drop net boats eventually replaced them.

Floyd Farrell (L) and Otto Marks lifting a drop net out of a rowboat.

Another boat used back then was a boat referred to as the “gasoline boat”. It too was used to tow a rowboat. They were nothing more than a flat bottom scow with no cabin and a center-mounted engine. The term was probably used to differentiate between the towboat and rowboat.

Leonard Dombrowski gassing up a “gasoline boat” at the mouth of the Suamico River.

Perch

The Bay of Green Bay has always been known as a hot bed for Yellow Perch or Bay Perch as they're referred to around these parts. They've been the staple for Friday night fish fries since Lord knows when. They have a distinct taste and the locals know if they're getting fed Bay perch or perch brought in from Lake Superior or Erie. They just taste better!

But due to over-fishing, mismanagement, nature, pollution, or whatever, their numbers diminished to a point where the DNR (Department of Natural Resourses) needed to impose quotas for both sport and commercial fishing. The quotas, along with attrition, took their toll on the number of commercial fisherman out of Suamico, from a high of more than 30 boats at one time to fewer than 5 today. During the 1950s when I fished there were about 15. The quota system began in 1983.

The good news is the perch are coming back but I doubt whether Suamico will ever be the fishing mecca it once was.

Yellow Perch of various legal sizes from a day's catch.

The Daily Routine

Being born to a fishing family, it was only natural I'd be working the boats as soon as I was strong enough to pull the anchor line. My Dad quit fishing full time around 1950 and got a job in Green Bay at Zoll Stone. He bought a fishing rig and partnered with Art and Bill Devroy to run it. They hired Rollie Belschner to work full time on the boat with Bill. Art worked the twine yard. I worked with them during the summer months.

Rollie Belshner and Dan Williquette. (Circa 1955)

The day usually began around five a.m. We lived next door to the Devroys so Bill would wake me up every morning by knocking on my bedroom window. From there it was pack a lunch and head for the dock.

The first thing we had to do was get ice from the icehouse. Most fishermen had an ice-house. (See Making Ice) In the winter we'd harvest enough ice from the river to last all summer. We'd dig a chunk out of the sawdust, rinse it off in the river and chink it up. The ice and a wet burlap would keep the days catch fresh. Eventually the fish houses had ice-making machines and we'd get the ice from them. They even had a machine that would chink it.(crush it.)

Before we'd leave we made sure we had everything; check the gas and oil, fish boxes, pike pole, rubber pants and boots, shovels, drinking water, our lunches and canvas gloves.

But first, if the wind was blowing, we'd jump in the car and take a ride down to the mouth of the river to see how rough the Bay looked. If it was too rough we'd stay home and work in the twine yard or just screw around all day. We always found time for fun. If it didn't look bad we'd head for our nets.

To find our nets we didn't use a very sophisticated navigation system. We'd use landmarks on the shore. Our system consisted of lining up the square light with Point Sable, or the shiny roof on the east shore with the power plant or the black can with Point Comfort. It was my job to spot the buoys once we'd get close. If it was foggy, we stayed home. Some guys used dead reckoning with compasses and timers. A certain heading for "x" amount of time at "x" rpms would bring them right to their nets.

Some times it would get foggy when we were out there. That created a problem. We'd kind of have to feel our way home and rely on instinct. Every now and then we'd have to stop the boat, shut it off and listen for other boats in the vicinity. If you didn't hear any, you'd holler in case someone else was doing the same thing. It wasn't unusual to have someone right next to you and not even realize they're there until they holler back and scare the crap out of you. You'd also listen for the bells on the two channel buoys at the mouth of the river. Once we were coming in and I could hear the bells and thought I was right on track. When all of a sudden we started hitting bottom. Coming out of the fog about thirty feet ahead of us was Nora Rice's cottage. It came out of the fog like a ghost. It was about a quarter mile south of the river on Long Tail. We missed the river, but at least we knew where we were.

Occasionally we'd get caught in a summer storm, or squall. The Bay can get really rough in no time at all. Being out there at the mercy of the lighting and wind can get very scary and there's no place to hide. Once we were trying to get home during a storm but it got so rough we were at risk of rolling over. DeWitts were about a hundred yards behind us. Their boat would roll so much I could see the whole underside, the keel, prop and rudder. We both decided to throw anchor and ride it out. Throwing the anchor puts the bow into the wind and keeps the boat from getting sideways.

A nor'easter was the fishermen's worst nightmare. The wind blowing straight down the Bay could build huge waves causing the nets to roll and render them unfishable. They would need to be pulled and reset. Minor twists could be rectified without pulling. Often the weather condition would last several days before we could get out to the nets to repair the damage, by then the fish would all be dead from being tossed around. The term used to describe the condition was, the nets were "knocked out."

We were heading over to the east shore one day to lift a gang of nets we had set between Point Sable and the Fox River. When we got there some of the pots were so full of dead fish they were floating on the surface. It takes a lot of dead fish to float a net off the bottom. We figured some business on the Fox River dumped something they shouldn't have and caused all the fish in the south end of the Bay to seek cleaner water. Those that ended up in our nets were trapped and died. We just opened the ends and dumped the whole pot.

On a lighter note, we'd listen to the radio while we sorted fish. Rock and roll was in its infancy then, and we'd listen to WAPL and DJ, Bob Bandy. He'd play everything but Elvis. I don't know why he hated Elvis; everybody else loved him, except our parents. The music from "My Fair Lady" was very popular too.

After we were done lifting, we'd have to wash the floor down. The fish slime made it pretty slippery. All the water would collect in the bilge. The automatic baler was a pipe that ran from the bilge out the bottom of the boat. Once the boat was moving you'd remove the pipe plug and the water would be siphoned out. When the bilge is emptied you're supposed to put the plug back in. If you don't, the water comes back in the bilge as soon as the boat comes to a stop. Guess who forgot to put the plug back in once? We got in around noon and I went home

and was eating lunch when it hit me. I ran back as fast as I could. When I got there it was about 6 inches from going under. I felt around and found the plug and got it in. Then I started bailing with a 5-gallon bucket. I was so happy to see it start coming up.
I've seen boats with just the exhaust pipe sticking out of the water. Believe me they're not easy to get floating again.

Like I said, we always found time for fun. We'd play pranks on each other all the time. Dick Devroy was coming up the river one day and Lyle Kirby, my brother Mike and Curt Nelson were waiting for him on the County J Bridge. They waved, and he waved back. As he passed under the bridge, they ran to the other side and rained buckets of water on him. He never looked up, never looked back, just kept right on a going as if nothing happened. But he was already scheming on how he'd get them back.

The word was, "never trust a Suamico Boy with a water hose or get between him and the river", chances are you're gonna get wet.

I owed Dick and his nephew, Johnny Brickner, a farm boy from Pulaski, for pushing me off the dock. As I got off the school bus one cold October day I saw Dick sitting on a fish box mending nets in the garage. I went down the basement, got an ice-cold bucket of water, walked over to the garage, threw open the door and by the time he looked up to see who it was, the water was on the way. The next summer his nephew showed up at Dick's with his dad's car with his girlfriend sitting next to him in the front seat and another couple in the back. They'd just come back from swimming down at Sunset Beach. I saw them talking to Dick with the car windows rolled down. I went down into the basement and got an ice-cold bucket of water; you know the rest.

On my 30th birthday we were going out to my brother Mike's for a cookout, or so I thought. He lived in the house we grew up in, in Suamico. When we got there Dick, Lyle, Mike, Curt and a few others grabbed me as I stepped from the car, carried me down to the river and in I went. It was May and the river was cold but I was more concerned about stepping on broken glass from the hundreds of bottles we'd shot with .22 rifles over the years. My wife, Sandy, was in on it too; she brought a change of clothes for me. I owe her one.

We not only shot bottles in the river, we'd shoot anything that got in our sights. I was shooting at a duck that was sitting down the river from me and being young and stupid and not knowing any better, I failed to realize that .22 bullets have a tendency to ricochet. I heard Donny Campbell come out of his house, screaming at the top of his lungs, "STOP SHOOTING – STOP SHOOTING". He jumped in his car and drove right down to where I was by the river and said the bullets were coming right through the side of his house. They almost hit grandma Laura Valentine, Mabel Campbell's mother.

Another time the .22 got me in trouble, Bill and I and Rollie were lifting north of Little Tail when our boat, the "Isle of Capri," broke down. We weren't far from the Little Suamico River so we used our scoop shovels to row in there. We walked into town and called Dick to come get us with the "Arbutus" and bring the .22, there was a pond there full of bisamaroons* (bullfrogs). We didn't know who the pond belonged to and we didn't care, all we wanted were the frogs. All of a sudden Eugene Marks was standing on the other shore telling us to stop shooting. It was his property and the bullets were ricocheting and almost hitting him. That was it for the .22.

*We always referred to bullfrogs as bisamaroons. But I'll be damned if I can find it in the dictionary for definition or spelling. Maybe it's just a Suamocian word.

The Nets

The nets we fished with were called fyke nets or drop nets. They were also referred to as double headers, meaning they had a pot on each end of the lead. They consisted of three distinct components: the pots, where the fish ended up; the lead, which directed the fish to the pots; and the hearts, a transitional section between the lead and pot which didn't allow the fish to back out and kept them swimming into the pot. They were set in traditional places known locally by such names as north of Little Tail, the frying pan, Red Banks, the east shore, the black can, outside of Long Tail, all productive fishing holes.

The business end, the pot, where the fish end up.

Once a pot was lifted, the fish needed to be sorted. This was accomplished using a scoop shovel with a gauge fastened to the handle. Perch were either regular, 7-1/2 inches to 8-1/2 inches; jumbos, 8-1/2 to 10-1/2 inches, or slabs, over 10-1/2 inches. Rough fish, suckers and carp, were sold to the mink ranches for mink feed. They didn't pay much but usually paid for our gas. You'd need to catch an awful lot of them today! We also kept walleye and northern if they were long enough.

In the picture of Arnie and Herb Zakowski on the next page, you'll see they had the ideal combination for sorting fish; one sorted right handed and the other left. It was cumbersome for two righties or two lefties to sort at the same box. It worked out that way for Bill and me too. Luckily it came natural to me. It was one of the few things in life I did left-handed.

Hauling in a pot full of perch.

Arnie (L) and Herb Zakowski sorting perch.

A bonus was the Bay crabs that would be all over the nets eating the dead fish. We'd save them in live boxes by our dock. And when we'd have about 500 we'd have a crab cook. You'd have to pick a day when the wind was just right so people driving by couldn't tell you were cooking crabs. Otherwise they'd stop and want to join in and there were never enough to share. If you've ever had crabs cooked in salt and dill you know what I'm talking about.

Teresa and Dick Devroy, my brother Mike Williquette and his son Ben, brother Jim and daughter Shae, and my wife Sandy.

Max

The Devroys had a big black dog, named Max. He weighed over 100 lbs. He'd sometimes follow us down the river when we'd leave in the morning. The mouth of the river was as far as he could go. Once one of the other boats picked him up swimming in the Bay outside the river, probably trying to follow us. Sometimes he'd be there waiting for us when we'd come in. So we'd pull the boat over to the shore and give him a ride home.

Larry Moore had dogs tied by the river and so did Ed Coopman. They'd start barking soon as they saw Max. He was half blind but he knew they were there. Out of the boat he'd go. He'd come limping home all chewed up. He came home full of BBs once. Someone found him in their living room with their lady dog. He went right through the screen door.

I used to dread having to go next door after dark to get Hermie Devroy* to come to the phone (they didn't have one). As soon as I'd leave our front door I'd start calling Max's name to let him know I was coming. He slept right by their back door or in their entrance. Soon as I'd get around the corner of their house, I'd start taking real short steps and breaths, hoping he'd heard me and knew I was coming. I could feel the hair on my legs standing up. If he was sleeping and you'd startle him, that was the worst fear. He was half deaf too. He never bit anybody, but you could never be sure. If he wasn't by the back door, now you had to negotiate the entrance. The worst that was going to happen is you were going to get the living daylights scared out of you. You knew it was coming, you just didn't know when. It was as inevitable as Lucy pulling the football up on Charley Brown.

Dick and I had two little dogs named Flash and Buck. One day they were raising a ruckus in the twine yard, barking like crazy at a young shy poke** they had cornered but were afraid to get near. Max heard them and came running. He ran right between the two pups and right over the bird, sending it tumbling. He couldn't see the shy poke even as he came nose to nose with it. The shy poke made a mistake by pecking Max on the nose. End of story. End of shy poke.

Ronnie Schiltz accidentally hit Max with his car one night, and killed him. It practically demolished his brother Jim's 1950 Ford. Bill and Dick dug a hole out behind the house and buried him. When they put him in the hole, his leg was twisted underneath him and Dick didn't like it. Bill went down and straightened it out. They wanted him to be comfortable.

*Hermie Devroy Was Art's Wife and Bill and Dick's Mom.
**A black crowned night heron.

Lyle Valentine's Fish House.

Once we were done lifting, we'd head for home. We'd stop at Larry Moore's mink ranch with the rough fish and then to our dock about a quarter mile west of the County J Bridge. We'd load the fish on the truck and head for Lyle Valentine's Fish House. We'd weigh the fish and get a receipt. Payday was once a week. My pay was the proceeds from one double header drop net.

While there, we'd tease the girls a little. Once Dick saw a bottle of soda in the cooler and came in where the gals were cleaning fish and announced, "who's ever soda's that was, it sure hit the spot." He was only kidding, but Doris Brunette didn't think it was so funny. She chased Dick screaming she was going to "cut his heart out." I think she was only kidding, but Dick wasn't taking any chances and didn't let her catch him. For a big gal, she sure could run. Dick was praying she'd tire out before he did.

Lyle Valentine, "Slicker" as he was known, and his family lived above the fish house. He had one of the first TV's in Suamico. I remember going up there with my Dad and watching "snow" and an occasional picture from a Milwaukee station. A short time later, we got one. By then you could sit and watch WBAY's test pattern for hours.

Gokey's Garage and Lyle Valentine's Fish House in the background.

Lyle later rented the upstairs to a bunch of guys from Upper Michigan trying out for the original "Green Bay Bobcat Hockey Team." Ken Rouhonen, Roger Maki and Ron Katlina are a few names that come to mind. We got to know them pretty well and had lots of good times with them. They introduced us to pizza and pasties.

Ken, or Kenner as we knew him, drove Gokey's stock car. We chummed a lot with Ron Katlina. He had a brand new 58 Chevy Impala. Darold Rogers and I stood for his wedding.

Gokey's Garage.

From there it was to Gokey's, a Standard Oil station on the corner of Velp and County B. Ralph was a good old guy who everybody liked, I guess because he tolerated us. The first thing we'd get was a big bottle of RC Cola and a bag of Planter's Peanuts, pour the peanuts in the cola and sit around and talk smart. Johnny Bender was the best joke teller and Dick Devroy was the best storyteller.

The pop machine wasn't very well grounded and you'd usually get a shock if you weren't careful getting your soda. It was fun watching unwary strangers reactions. Who was gonna warn them? A wet cat came in out of the rain once, rubbed against it and fell over dead.

Once the Gokey boys, Allen and Dick, put up a sign " gas-$.17 a gallon," and in smaller letters "plus tax." That would have brought the price up to the going price of about $.25 a gallon. (Hard to believe). Cars were sliding sideways down the highway trying to stop to get some cheap gas. Velp Avenue was Highway 41 and 141 at the time.

I remember when Ralph died. It was on a Sunday and we were sitting in St. Benedict's Catholic Church waiting for the service to start. It was unusual for Father Betley not to be on time. Turns out he was at Ralph's home giving him the Last Sacraments. He died unexpectedly. When he announced what had happened, everyone was shocked. I was pretty sad. I really liked him.

Ralph Gokey, Joe Komorowski, Ed Tilque, and Dick Gokey. Joe Komorowski later opened his own Sinclair gas station a few blocks north on then highway 41/141.

Maintenance Of The Nets

It wasn't all fun and games though. The nets in those days were made of cotton and required a lot of maintenance. Holes were a constant problem due to chafing especially in the pot. A hole in the pot could cost you much of the catch. The pots would be thoroughly inspected after each lift and holes mended before they were set back.

It wasn't unusual to lift a pot and find it empty with a big gaping hole staring at you. Besides chafing, holes could be caused by a large game fish chewing its way out or unintentionally by a sports fisherman. Sport fishermen sometimes think if they see a commercial fisherman's net in the water that it must be a good place to fish, and often end up getting their gear or anchor tangled in the net

Arnie and Herb Zakowski looking for holes before resetting the net.

Before the nets were set each year they needed to be tarred, and once during the summer they had to be taken out of service to be tarred again. Before they were tarred they needed to be cleaned and dried. Moss and mud would cling to them while in the water and needed to be removed. To clean them they were washed with a high-pressure water pump. The pump sat on the dock and was driven by a Model "A" engine. The nets were pulled out of the boat across the dock where the washing would take place. As they were being washed they were also being loaded on the twine yard truck and taken to the field to be spread out to dry.

Mark (L), Frank and Bob Maricque washing nets that were just pulled from the bay to be tarred.

Herb Zakowski checking the hearts for holes before tarring.

Once the nets were spread in the field to dry, pipes would be driven in the ground and the top poles of the hearts hoisted up on them and tied to open the hearts to facilitate inspection and mending. This is the way they look when in the water. Of course now instead of catching fish they'd often catch birds. We'd have to chase them out of the end of the pot.

Leaky floats would need to be replaced. The floats are what give the net its shape in the water. They are tied about three feet apart all along the top lines to float the net into shape. Without them the net would lay flat on the bottom. They are nothing more than a sealed galvanized tin can. Over time they develop leaks. Iron weights are tied to the bottom lines to hold the net down.

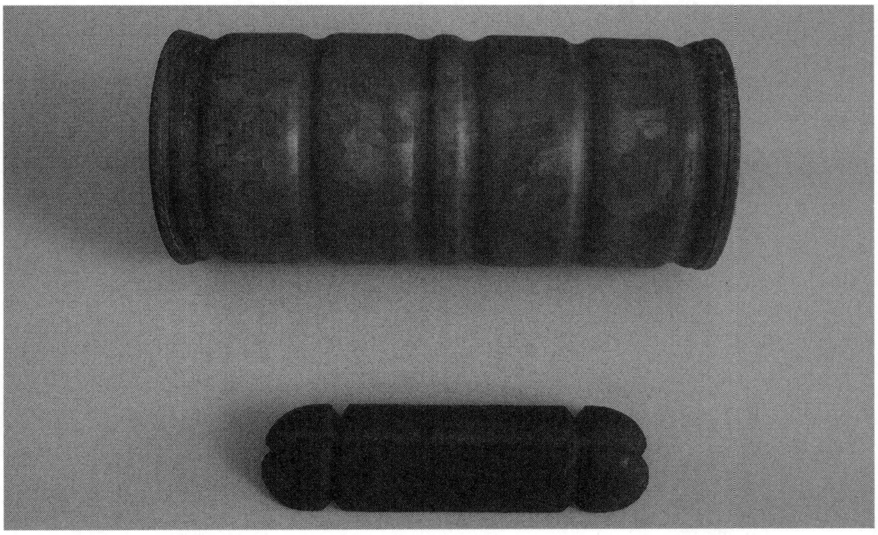

The float on top is 7" long by 3" in diameter. The iron sinker is 5" long.

Gary Rhodes removing the bottom spreader poles.

Before tarring, the nets had to be disassembled first. The lead would be separated from the pot and hearts. The pots and hearts are a single unit. The spreader poles, which give the hearts their shape to make them functional, needed to be removed. The top poles were usually 2" X 2" redwood or cedar and the bottom poles 1/2 inch galvanized pipe.

Tarring was the method used to protect and preserve the nets. Tarring began by picking a nice warm sunny day and starting a wood fire under the tar kettle to get the tar boiling. The smell of the boiling tar is firmly imbedded in anyone's memory from that era.

John Valentine (left), and Norm Williquette taking a net from the tar kettle and putting it on the stone bolt. A stone bolt is like a sled with wooden runners to be towed on grass. The kids are Dan Williquette and cousin Joan LeFebvre.

Ice cream also goes good on a warm summer day.

The hearts and pot are tarred separately from the lead. The different sections are immersed in the boiling tar and cooked, then raised up and allowed to drain. From there they're once again spread in the field to dry. Once dry, they are reassembled and reset to fish again. After time the fishermen abandon the round tar kettles for ones large enough to immerse the entire net, poles and all, thus eliminating having to disassemble them every time.

The cotton eventually gave way to a synthetic fiber called "Nycot," a combination of nylon and cotton, and required far less maintenance and fewer holes for the fish to escape. Tarring was reduced to only once a year, before the season started.

Most fishermen assembled their own nets from twine purchased from local suppliers. Many bought from The Linen Thread Co. in Marinette. It came in different gauges or weights. The lead and hearts were made of 12 thread and the pots of 18 thread. Each thread was made up of three strands, and each strand was made up of a number of threads; 12-thread was three strands, each containing four threads. 3 x 4 = 12 thread.

You could buy the twine as woven netting or spools of thread for those who preferred to knit their own nets from scratch, a long arduous project. Bill Devroy was one who preferred knitting his own. In the summer he'd sit out under the apple trees and knit and in winter up in his room. A light in his upstairs room would cast a shadow of him knitting onto our garage door.

When we'd make our trip to Marinette for twine, we'd also slip into Menominee, Michigan and load up on oleomargarine. The only oleo you could buy in Wisconsin came in a one-pound plastic bag. It was white and to make it yellow, there was a bead of yellow dye in the package. You needed to break the bead and then knead the dye into the oleo to make it look like butter. Wisconsin was the dairy state and if you were determined to buy oleo, and not butter, they made you work for it. It was an old "Blue Law" from days gone by. When the neighbors got wind you were going to Marinette, they'd put in their order.

Bill Devroy's Model "A" Ford was used as twine yard truck at times. At other times it was used for parades, all painted up in bright colors representing Suamico's VFW Post 9409. Many of the fishermen at that time were WWI, II and Korean War veterans.

Most twine yard trucks were built from Model "A" Fords and called "puddle jumpers." The body was stripped off from the windshield back and a wooden bed put in its place. Not ours. Ours was a made from a 1934, 4-door La Salle. We sawed it off just behind the driver's seat and made wooden truck bed from there back. It makes me sick when I think of what we did to that car; it'd be worth a fortune today.

Once we took it to Jenk's Channel to catch a beer. Before long it got dark out.
The area of the truck bed where the nets were loaded was an extension of the main bed and was lower and wider and stuck out about three feet on either side. It was only about shin high.

The toilets at Jenk's were outdoor privies and to get to them you had to walk between the cars. We were getting ready to leave when we found a guy lying on the ground behind the truck, moaning and clutching his shins hollering, "Look out for that truck!" Like I said, it was dark and he was probably in a hurry.

West Lakeside School and the Guild Hall

Back in the 1950s Suamico had two public buildings, West Lakeside School and the Guild Hall. Both are private residences now.

All future fishermen went to West Lakeside School. It was right in the middle of the fishing village. George Ladrow was across the street; Ben Valentine was on one side and John Valentine on the other. We were surrounded by twine yards and tar kettles.

Ladrow had an apple orchard we could raid, and we played ball on Ben Valentine's land. Everybody walked to school and carried their lunch. It was a three-room school with grades 1 through 8. Grades 1 and 2 were in the little room, 3, 4, 5 were in the middle room and 6, 7, 8 were in the big room. We took iodine pills and played with mercury. We got to ring the bell and raise and lower the flag. They added "under God" to the Pledge of Allegiance when we were kids.

The smell of boiling tar was a sure sign school would soon be out, along with the melting snow and the ditches full of pickerel, the fruit trees blossoming and the pussy willows. Summer vacation was just around the corner except for us Catholics; we had three weeks of Catechism to look forward to.

Our eighth grade class spent the first part of the year attending school in the Methodist Church basement, while the "new" Suamico Elementary School was getting some finishing touches. We were the first class to graduate from there and planted the first row of trees up the hill. It was 1955.

Graduation was held in Green Bay and my mother had bought me a new pair of shoes for the occasion. When it was time to go, we couldn't find the shoes. My dad was at work and Ma turned the house upside down looking for them. We ended up shining my old shoes. When the ceremony was over and we got home Ma told Dad what had happened. He knew right where the shoes were, in the bottom desk drawer! He said I was sleeping on the couch, got up, put the shoes in the drawer and went to bed. I don't remember doing it. Must have been walking in my sleep.

Behind the school was a field belonging to Art Fiest, a mink rancher who kept horses there that were to be slaughtered for mink feed. Tractors had replaced the workhorse and the no longer needed horses were being sold to the mink ranches. At recess the horses would come by the fence and we'd feed them grass. They were likable and easy to get attached to. The one I adopted had one blue eye and one brown.

We were also subjected to watching the horses being slaughtered; a practice I'm sure wouldn't be tolerated today. It was done with a .22 rifle shot to the forehead. We were fascinated by the fact we'd see the smoke from the gun, the horse fall, and then we'd hear the report from the gun. I guess it was a lesson learned on the relationship between the speed of light and speed of sound. So some good came out of it.

It's funny how certain things always come to the forefront of your memory when reminiscing. There are two things that always pop into my head, whenever I visit West Lakeside in my thoughts, and both involve Bobby Komorowski.

The first occasion was the day he started school in first grade. He didn't want anything to do with it. I can remember seeing his sisters, Marie and Victoria, dragging him through the schoolyard, one on each arm, and him with his heels dug in screaming in protest.

The second was a very snowy day when few kids showed up for school. We were in the "middle room", 3rd, 4th and 5th grades. I was in 5th, and Bobby was in 3rd. The teacher went to each grade and asked how many kids showed up for classes? There were some 5ths, some 4ths and Bobby volunteered, "there's only *tree tirds.*" We all talked like that, in our Wisconsin dialect, but that one really broke us up.

I'm guessing the picture was taken in the late teens or early 1920s in front of West Lakeside School. Many of the names are from fishing families; Lyle Valentine, Lavern and Charlotte Devroy are Dick and Bill's sisters, Sis Valentine is Bill Dickinson's mother, etc. Many of the others were from farm families.

First Row- Lyle Valentine, Ed Gritt, Irvin Valentine, Lloyd Massey, Peter Kalfas, Easly Clewly, Ruel Anderson, Zigmond Drzewiecki.

Second Row - Milton Sorenson, Tony Kozloski, Walter Posluszny, Neal Valentine, John Posluszny, Chester Gritt, Nelson Hanson, Dave Garsow.

Third Row - Lavern Devroy, Charlotte Devroy, Sophie Posluszny, Leona Diedrick, Beatrice Roderick, Elsie Brunette, Bertha Miller.

Fourth Row - Geraldine Farrell, unknown, Angline Dombrowski, Sis Valentine, (Teacher, unknown) Marion Devroy, Florence Massey, Viola Roderick.

The Guild Hall was the center of entertainment for the town. It hosted countless weddings and social events; Boy Scout Court of Honors, Cub Scout Blue and Gold banquets, political rallies, (I like Ike), square dances with John Gardner, dance recitals, banquets, you name it. It was even a roller rink. It belonged to St. Paul's Episcopal Church and the "women's guild" would prepare the food, mostly potluck.

The biggest attraction was "Cousin Fuzzy and His Cousins," and "Uncle Louie and the Town Hall Players," coming to town. This was an era before television so live entertainment was all there was. We'd get there early and sit up front on the floor. They'd put on a three-act play and then a dance afterwards. We'd wait in anticipation of the pistol going off, (blanks of course). It seemed in all the plays a jealous wife or husband would catch their spouse fooling around. All in fun.

After the play they'd put on a dance. The music was mostly polkas and waltzes. But our favorites were the "Bunny Hop" and the "Flying Dutchman."

All the little kids would run around and slide on the floor, young couples would dance and all the old men would sit around the edges, while their wives worked in the kitchen. It was very "Mayberry."

St. Paul's Episcopal Church's Guild Hall.

The grounds of the Guild Hall were also the community playgrounds. On the eastside was a huge oak tree with a rope hanging in it and on the other side was a swing set. The oak tree was the place kids liked to congregate and develop their social skill. It was the equivalent of going to the mall now days.

The river was our swimming hole. Behind Joe DeWitt's was the best spot. We used their garage to change clothes and spent countless hours swimming there. I'm surprised any of us survived; the river was also the town sewer. None of us ever got any illness you'd associate with swimming in all that pollution, but none of us got very *tall* either. (?)

For odd jobs during the summer, before we were old enough to work on the boats, we picked beans for Joe Destiche, split and piled cordwood for Jim Vickery, put up hay for Gordon Malchow, or pick pickles and strawberries for Evelyn Vickery.

There were also at least 5 mink ranches where you could pick up a job during pelting season. John Posluszny had one, as did George Lince, Art Fiest, Clayton Valentine, Larry Moore and probably others.

Many of the kids I chummed with back then had riding horses, Larry Valentine, Carol Coopman, Rosemary Moore and Dennis Peot all had one. I asked Larry Moore if I could use one for the summer one year and he let me use a quarter horse that had been brought to the ranch for mink feed, for reasons unknown. I named him General. He was very fast and all I had was a bridle and no saddle.

A funny thing about horses is they always run faster when headed for home. I was doing just that once when I got in front of Kozloski's. They had a dog that chased every thing that went past their house; cars, bikes, people and in this case a horse. The horse was on the dead run when the dog came out barking. I remember the horse making a sudden right move, me flying through the air, hitting the ground and rolling. Every time I'd roll, I'd see the horse getting further away until I came to a halt and he was rounding the corner by Massey's. As if that wasn't bad enough, now I had to deal with the dog! I limped home and when I got there, the horse was grazing in the twine yard as if nothing happened. I changed his name to "Mink Feed."

Fishing Through The Ice.

Fishing wasn't limited to just the summer months. When the ice started to form in the river and Bay, it was time to pull your nets out and start looking ahead to winter. Most winter fishing was done with gill nets, but some would set drop nets and pound nets through the ice.

Ray Tuttle (L), Donald Tuttle and Art Devroy lifting nets through the ice.

In order to get from the shore to where the nets were set, there was always an expansion crack that needed to be crossed. The crack was usually a few hundred yards off shore in 5 to 10 feet of water. And there was one on each side of the Bay and ran the length of the Bay.

The fishermen would all cross at the same place and had boards in place to act as a bridge. Sometimes the crack would expand to 4 or 5 feet, other times it would contract and push the ice so high you'd need to find a new crossing point. The condition could change drastically while you lifted your nets, from one extreme to the other.

Lyle Kirby and Mike Williquette were setting nets through the ice outside of Oconto with Dick Devroy one year. Mike and Dick were using a power drill attached to the front bumper of Dick's car, but the auger wasn't long enough to go completely through the ice, so Lyle was coming behind and finishing the hole with an ice chisel. The chisel got away from Lyle and he almost cut his toe off. They had to rush him to the emergency room in Oconto where they saved the toe.

A treat after freezing your butt all day out on the ice was stopping at Bing Moe's fish house in Oconto and having some freshly smoked chubs, still warm from the smoker. That made the day all worth it.

The mode of transportation those days was the snowmobile. Not the snowmobile we know today, but one fashioned from a Model "A" Ford. It was quite the contraption. It had an under carriage attached to the body, consisting of two additional wheels on each side. The forward tires had lugs that would mesh with a track that went around the two wheels and the rear drive wheels, which also had lugs. The center wheels acted as idlers. The front steering wheels were replaced with skis. It was very functional and served us well. In the dead of winter these thing sat outside on the shore near the trail to the nets. In order to get them started we'd light a can of fuel oil soaked rags and place it under the oil pan to warm up the oil in the engine so it would turn over to start.

When traveling, we always kept the doors open in case we'd break through the ice. Hopefully the doors would keep us from going all the way under. Thankfully we never had to find out.

To stay warm, we wore two pair of everything including long johns. The Model "A" manifold heater plus the doors being open didn't provide much heat. Lunch was always a hot thermos of cocoa and a big bologna sandwich. That would warm things up a bit.

Jimmy Tuttle tells of being in the back seat of his uncle Ray Tuttle's snowmobile when the back wheels broke through the ice while crossing the crack out of Peshtigo. Upon exiting the vehicle, "in a panic," he clambered over his uncle, pushing his face into the steering wheel, breaking his glasses and blooding his nose. All he could think of was the 65 feet of water underneath them and not wanting any part of it. What kept them from going under was the tongue of the trailer they were towing.

Dick Devroy standing on track of a snowmobile.

Making Ice

Making ice was another winter project. We didn't actually "make ice" it's just a term used to harvest ice from the river or ponds to use all summer to keep the fish fresh. The ice was cut out of the river right in front of the icehouse. It was cut into blocks using one-man ice handsaws in conjunction with a homemade power saw. The power saw was fashioned from a Model "T" Ford chassis, body removed, and one of the rear wheels replaced with a large circular saw blade.

Gene Vaillant (L) and Irv Sheedy getting ready to put the motorized ice saw to work.

The motorized large circular saw blade is making the initial cut. Chris Marks (L) and Jake Mutz are using one-man ice handsaws to make the final crosscut.

The blocks were floated up a channel right to the door of the icehouse and pulled in by a rope attached to a car bumper. Tiers were formed and the top tier would be covered with sawdust. It always amazed me how that sawdust could keep the ice from melting all summer. Those old timers were very ingenious in their ways.

We also used the ice for our iceboxes at home. It wasn't long before the refrigerator became a common household appliance. I guess you could say we grew up in the "ice age."

Ben, "Chief" Drzewiecki , guiding blocks of ice through the opening to the ice slide to be pulled up into the icehouse.

(L to R) Unknown, Jack Mutz, Paul Spotts and Tom Webster maneuvering a cake of ice inside the icehouse.

Jake Mutz working in front of the Mutz icehouse.

Icehouses were nothing more than a 20 x 20 foot wooden pole building with an opening toward the river or pond to get the ice in and out. Boards covered the opening and were removed as the tiers of ice went down. There were also wooden ladder rungs nailed to the outside to get up to the top. Some icehouses had roofs and some didn't.

Removing the ice from the river or pond created a very hazardous condition for skaters and kids playing on the ice until the hole froze over. Our mothers were constantly harping at us to stay away from where they were making ice. Precautions were used, Christmas trees were collected and used to mark the hole but it wasn't foolproof.

Dick Gokey learned the hard way. After a skating party on the carp pond behind his house, he needed to cross the river to get home. On his way, he stumbled into one of the trees marking a hole. He, the tree and the flashlight he was carrying all went through the thin ice and into the river. He managed to get out but his flashlight was lying on the bottom still lit. Standing there soaking wet and freezing, he was now faced with a decision; which was the lesser of two evils: leaving the flashlight there and showing up at home without it, or going back in the water after it? He chose what he considered the lesser at the time and went back in and got it!

Working In The Fish House

The going out and catching the fish was just part of the big picture of the Suamico fishing industry. There were the fish houses that processed the fish. At the time of this story there were three fish houses in Suamico. The fishermen had a choice of who to do business with. Bill Drzewiecki had The Suamico Fish Company, Lyle Valentine had Valentine Fisheries, and George Tuttle had Northern Fisheries. All processed fish for wholesale and retail. The fish were dressed, (filleted) by town's people, most of who were from fishing families, the wives and children of fishermen. You were paid by the pound, so the faster you went, the more you made. And they were fast. But if there were ever a contest, Violet (Budget) Nelson would win all the prizes. She was the best.

Picking up some extra cash during the annual smelt run. (L-R) Eugene Marks, unknown, unknown, Madeline Williquette, Jimmy Tuttle, Marie Valentine, Anne Mutz and Jake Mutz.

The catch was usually processed fresh but at times the fish would be frozen round, (whole) and processed when fresh fish weren't as available, usually in the winter months.

I got in trouble once with the ladies from the fish house. There was a cop sitting by the Methodist Church one day and I got to talking to him. I was probably 9 or 10 at the time. I told him if he wanted to catch speeders, wait until the fish house crew comes flying through here at noon. They were all in a hurry to get home and have lunch and get back to work. It got back to them and I got a lot of ribbing for that. I was pretty red faced but the ladies had a lot of fun with it.

(L-R) Lee Grages, Jerry Belschner, Dorothy Devroy, Annie Giese, Oliver Brunette, Julia DeWitt and Gladys Dombrowski.

(L-R) Anne Mutz, Jackie Massey, Margaret Keuken, Madeline Williquette, Zerlena Valentine, unknown. Anne, with Jake's cigar in her mouth, hamming it up for the camera.

Smelt Fishing

In the spring of the year attention shifted to a different kind of fishing. The annual smelt run would begin around the first of May. After a few warm spring rains, the streams would open up and the smelt would seek them out to spawn. The run would start on the Lake Michigan shore first and move around to the bay shores. On the Lakeside we'd set north of Algoma at Stoney Creek, and on the bay side at Namur.

Pound nets (commonly pronounced, pond nets) were used for fishing smelt. The smelt swam along the shoreline. The pound net's lead was attached to the shore and ran perpendicular from the shore out to the pot, which was set in about four feet of water. Stakes were driven with a hand held pile driver to support the net.

(L to R) Mike Williquette, Lyle Kirby and Tommy Valentine tend to their pound net.

The fish ran at night so we'd have to stay awake and check the nets about once an hour. Once the run started, the net would fill up in a hurry. When it was full we'd push a rowboat out to the net and start scooping them into the boat. A log was used to roll the net to keep working the fish to one side. When the boat was full, we'd push it to the shore, shovel the fish in boxes, and load them on a truck. When the truck was full, we'd head for home.

Wallygoo Bill Dickinson was heading home with a truckload one morning, after fishing all night, when he fell asleep and rolled the truck over in a farmer's field. The farmer got a free truckload of fertilizer.

If I wasn't catching smelt, I'd be cleaning them. I'd get off the school bus at Lyle Valentine's fish house and clean smelt till 9:00 p.m. You could always tell where kids were from at Pulaski High School. The Suamico kids smelled like fish and you know what the Pulaski kids smelled like! Just kidding. (Pulaski's a farming community.)

Dick Devroy came over to visit Lyle where he had his nets set on the east shore. He brought along a gill net to set off one of the stakes that support the pound net. (Strictly illegal.) He rowed out into the night in Lyle's aluminum rowboat, to a few hundred feet off shore. Shortly, some bystanders came by and were talking to Lyle about what he was doing. About then Dick came rowing in out of the night. They asked, "Where did you come from?" Dick said "Oconto - and are my arms tired." (Oconto is on west shore of the bay, about 10 miles away.)

My brother Mike and I and my brother in law, John Van Lieshout, took a ride over to Namur one night just to see if the smelt were running and maybe catch a few if they were. There happens to be a tavern there so we went in to see if there were any fishermen in there we knew. Sure enough, Pete Ertel was in there waiting for his crew to show up. He had a pound net set just down the shore from the bar.

We were sitting with Pete, having a few beers, when this bunch of guys came in. We asked Pete if that was his crew? Pete said "hell no, they're too clean cut, my crew is grubby." Pete was known to frequent Broadway in Green Bay and that's where he recruited his crew. (Broadway used to be the bad part of town).

It wasn't long and a big red convertible came sliding in the driveway with the wildest bunch of characters you've ever seen, hanging half in and half out of the car. Pete said, "That's my crew!" He was right, they were grubby. It was obvious they'd just come from Broadway too, and were feeling no pain.

He got them organized and they started pushing the boat out to check the net. We were sitting on the shore watching. Pete was standing in the bow looking like George Washington crossing the Delaware.

About three quarters of the way out to the net, they realized the boat was sinking. It was kind of rough and the boat was down to just the gunnels sticking out when they tried to retreat to the shore. By then the fish boxes were floating out to sea and Pete was screaming "get me back to shore, the captain's not supposed to get wet!" His crew had all but abandoned him trying to keep from drowning themselves. Pete got wet, they all survived and we had the best laugh we ever had in our lives. We still laugh about it when we get together. It was a lot funnier being there than trying to explain it.

Lyle Kirby with a boatload of smelt

Gill nets

Gill net fishing was another type of fishing done out of Suamico. Most everybody fished drop nets back then, but there were a few gill net boats in the river. Most gill netting took place later in the year when the weather turned cold. Gill net boats are totally enclosed and have a stove to provide heat. Today, most everyone fishes gill nets.

Gill nets work just as the name implies. The net is a long lead and so fine it's invisible to the fish. When the fish encounter it, the small ones swim through it, the big ones swim away when confronted with it, but those just right swim into it up to their gills and they're caught. The mesh, or size of the holes in the netting, is sized to catch only legal sized fish.

The whole net is pulled into the boat by a winch and the fish removed once inside the boat, out of the weather.

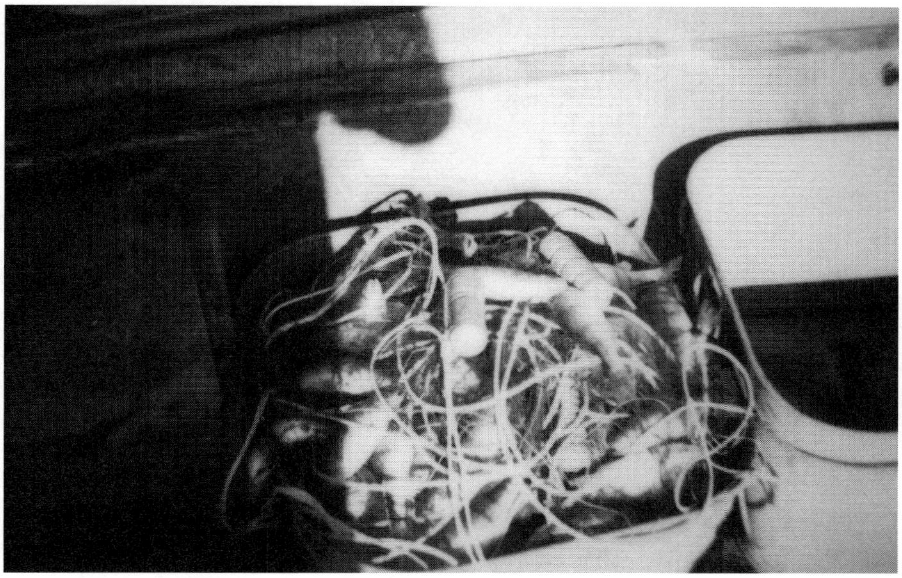

Whole gill net, fish and all, pulled into a tub to be picked later.

Gill net boats parked at Bender's Landing

Before synthetic twine was used, Gill nets were made of cotton and needed to be dried when they were out of the water or they'd rot. Here gill nets are wound on drying racks.

Crabbers

There are those who made a pretty good living crabbing. Jake and Jack Mutz crabbed for more than 40 years on the Bay. Tony Kozloski and his family did the same.

The crabs were sold to local taverns, but some were shipped as far as the East Coast. Locals who wanted a couple hundred could also buy them. You could either buy them cleaned or clean them yourself. Most opted for the latter.

Jake Mutz soaking up a few of his hundreds of crab boxes for the upcoming season.

My brother Mike and Tommy Valentine went to Jake Mutz to buy some crabs one Sunday afternoon. Jake was still in his suit from attending church. On his way home he stopped to pick up the Sunday paper, either at the White Front Tavern (it's gone now) or Walsh's Tavern (now the Four-Way Bar). I think one had the Milwaukee Sentinel and the other the Milwaukee Journal; there was no Sunday Green Bay Press-Gazette at that time.

He and the boys walked out on the dock to where the live rack was kept and Jake grabbed the rope and began pulling it up. But instead of the box coming up it got caught on the edge of the dock and pulled Jake right into the pond. He made a summersault, landed on the bottom in about three feet of water and was lying there looking up at the boys with his ever-present cigar still in his mouth. He came up laughing; wanting to know which one had pushed him in, and if they had a light for his cigar?

Jake (running the motor) and Jack Mutz lifting crab boxes

Crab boxes were generally about 9" X 9" X 18" with a tunnel on one end for the crabs to enter. They were made of cedar laths with a bit of concrete to give them weight. A float was attached to locate and identify them. They were baited with chunks of carp or suckers.

Bill Devroy made and sold hundreds of crab boxes as a sideline. Dick Devroy set boxes on Chute Pond for his own consumption and also kept me supplied. My cousin Jimmy Nier crabbed for years on Lake Metonga. My brother Mike and Lyle Kirby crabbed in Eagle River, as did Horace and Clara Drzewiecki.

Years ago taverns would give them away to draw business. Today you pay dearly for them.

And those Cajuns can say what they want about how to prepare crayfish, I'll take the way we cook them anytime over their fancy spices. Ever tried theirs? You can't eat um! Give me plain old canning salt and some fresh dill out of the garden anytime.

Seiners

There were some fishermen who only seined carp for a living. It was a huge operation that involved digging ponds to hold the fish until the market price was right or for certain religious holidays. Chicago and New York were big markets for carp. The Hermes family was a major carp seining operation until the FDA determined the PCB and mercury levels were too high for human consumption. Brothers, Jake and Jack Mutz were also big into seining and had their own carp pond on their property north of Long Tail Beach.

Jack and Jake Mutz returning to port with their seine and live racks in tow.

The fish were caught in the Bay, usually in the sloughs inside Long Tail and Little Tail. They were seined and transferred to live racks then towed to the ponds where they were released to swim freely until shipping time. They were either shipped round (whole) or live in water filled train cars or semi trailers.

Several ponds still remain in Suamico as are evident on the plat map in the beginning of the book. They were dug adjacent to the river. A sluiceway was dug between the river and the ponds and the water diverted to supply the ponds with fresh water. A cement abutment with steel grated gates on either end let the water flow through and kept the carp contained. The sluiceways are also where the carp were transferred from the live racks to the pond.

Some ponds had aeration pumps that would keep the water oxygenated when the rivers were stagnant. The one behind our house didn't have an aeration system because most of the time there was adequate flow from the river. But there were times during the dog days of summer when the pond would get depleted of fresh water and there would be a big die off. A lot of dead carp can give off an awful stench.

Jake Mutz getting things lined up for a pull.

Jack Mutz pulling a seine by hand.

Bill Dickinson, Gerry Shallow and myself seined inside Long Tail one year. Gerry and I sat on the shore with our heels dug in and pulled the seine by hand. Billy ran the boat. We weren't as mechanized as some of the bigger operations. We didn't utilize a carp pond either; the fish went directly from the live racks to Mike Buckarma's semi. We pitched the carp, a term used for throwing the fish by hand into the truck.

The only nice thing I can remember about seining is we weren't far from Harbor Lights Resort at lunchtime. We'd go in there and have a burger and a couple of beers. I wasn't quite 18 at the time; this was early May and my birthday wasn't until the end of May. When my birthday finally rolled around I announced to Tony (Drzewiecki) that I was finally old enough to be in here. He said I'd been rushing the season before, and then he bought me a beer.

Then there were those who seined minnows. Stanley Benedyk, Leona Mealy and Frank Kalfas also had ponds to store minnows in. Their ponds were located east of the Chicago/Northwestern tracks across from what is now Vickery Village.

Peter Devroy owned the carp pond behind our house. His grandson, Bill Devroy, recalls his grandfather feeding the carp. He'd pound on a pail and the carp would all come to the west end of the pond and he'd throw them corn. The water would roil with carp.

As kids, we put the ponds to good use in the winter. We'd shovel an area big enough for a hockey rink and have bon fires and skating parties. Every time it snowed we'd have to shovel it before we could get a hockey game going. The ice was always smoother there away from the river currents. As soon as it froze over, we'd be on it. We called it "rubber ice" because it was very thin and kind of floated on the water until it froze solid. (If our mothers only knew.)

A Saturday of sliding down the hill. On sled on left is Mike Williquette and Cathy (Roderick) Johnson, standing behind is Dan Williquette. On sled on right is Helen (Nelson) Shallow and Curt Nelson, standing is Virginia (Nelson) Kamps.

There was a hill on the way to the pond where we'd sled if we weren't skating. We'd sometimes haul water from the river to ice down the hill. A good run would take you down the hill, through the swamp, through the woods and on to the river. Then do it over and over and over again. Endless days and endless energy.

The "Isle Of Capri" can be seen sitting on the bank in the background in the above picture. It was pulled out for the winter and needed to be tethered because every spring, when the river broke up, the ice would jam up under the old iron bridge over County J and all the low lands west of the bridge would flood. If the water got too high it could float the boat off it's mooring. The county crews would use dynamite to break up the ice to get the river flowing again. The iron bridge near the Catholic Church was once lifted off its footings by an ice jam and moved 40 feet down stream.

The old iron bridges would rattle when a car would cross them. That noise, plus the steam engines spinning their wheels on the railroad tracks, and the thousands of Whistling Swans that would safe harbor at the mouth of the river on their migration route, are some of the sounds that would lull you to sleep at night; along with the frogs and crickets.

Fish Boxes

Wilfred Rosera was the man who supplied the town with fish boxes. He made them out of salvaged fruit and vegetable crates. They were very sturdy and served both the fishermen and fish houses well. They held 80 lbs. of perch or 100 lbs. of smelt. He built them all by hand in a garage behind his house. They were affectionately known as "Willie Boxes." He had plenty of help from his grandkids, Randy and Jimmy Lewis and their buddy, Curt Nelson.

Wilfred was a very proud man. Once my Dad and some other fellas and Wilfred were going to cut some tamarack trees. They were used for pound net stakes. They all met at Gokey's garage. Each man was bragging how sharp his axe was. My Dad looked at Wilfred's axe and asked him if he was going to saw the trees down with it. Wilfred got mad and went home. No apologizing could change his mind.

"Willie Boxes"

I asked Curt if he had any stories to share about working for Wilfred. He told of how they took the salvaged crates apart, carefully pulling out the nail so as to not split the boards and how everything was nailed back together by hand, no nailing machines. He also told of kegs of different sized nails for different areas of the boxes.

One particular keg really caught his attention; he reached in to grab a handful of nails from this seemingly full keg and discovered he could only get his fingers in a couple of inches before his fingers touched bottom. He called Jimmy over and showed him. The two of them stumbled onto Wilfred's secret money stash. The keg had a false bottom and in it were bags of each denomination of coins and coffee cans of folding money. After some consideration the two 14 year olds thought better of seeing if Wilfred would miss a few bucks. Fear and common sense and good upbringing won out. Good thing. Next day the keg was gone. Somehow Wilfred knew someone was in there.

The Long Tail Point Light Houses.

A lone sentinel remains standing today to steer the ghosts of sailing ships safely past the point of Long Tail. Though the oldest of its contemporaries it is all that remains of the 3 lighthouses constructed over the centuries on that finger of land. Built in the early 1800s out of fieldstone, it is affectionately referred to as the "Old Stone Light." In the 1950s there were remnants of the building that housed the "keeper" and his family next to it. As kids we'd sometime camp out there over night.

The original "Stone Light" viewed from the Long Tail slough.

In the mid 1800s the second lighthouse was built just to the north of the Stone Light.

The dock, where people and supplies were delivered, was 600 to 800 feet long and ran from the slough, across the marsh to the lighthouse.

The second lighthouse built to replace the Stone Light, was a two-story structure that housed the living quarters of the keeper, the assistant keeper and the tower where the beacon was housed. The families of the keepers would spend summers on the island and return to the mainland when school started. The keepers would stay until the navigation season ended.

Joan Mutz recalls her father (Jake) telling of the lighthouse keeper, Andrew Weber and his family, traveling from Green Bay to the Mutz property north of Long Tail Beach, and that he and his brother Jack would boat them and their supplies from the mainland to the island. They would make the trip several times during the navigational season.

The "Square Light."

In the late 1800s the third lighthouse was constructed several hundred feet off the point of Long Tail nearer the shipping channel. It became known as the "Square Light," for obvious reasons. The keepers remained living in the second lighthouse's living quarters while they maintained the Square Light.

Maintenance consisted of walking out to the end of the island about a half hour before sunset, row out to the lighthouse and light the kerosene lantern that provided the light for the beacon. The lantern operated on the same principle as a Coleman lantern: kerosene, compressed air and a mantle, plus powerful lenses to magnify the light. They'd spend the night out there and several times during the night they'd check to make sure the light was working properly. If fog rolled in, they'd activate the foghorn, which would send out a blast once a minute. In the morning they'd extinguish the flame, clean the lenses and the windows of the crib, the glass enclosure that surrounded the lantern, and then return to their quarters on the island. This continued up until around 1940 when the outer light, or entrance light, was built and the keepers were then transferred out there.

When the second lighthouse was abandoned, it was purchased by the Malchow family, disassembled and moved in pieces across the ice to their property, south of Long Tail Beach. An attempt to move it across the ice intact failed due to ice conditions and insurance liabilities.

The Square Light lasted until April of 1973 when a ferocious northeaster toppled it. It was such a devastating storm and flood that the Federal Government declared the City of Green Bay and the surrounding communities on the south end of the Bay a disaster area. Today, automated buoys line the harbor and are tended to by the U.S. Coast Guard.

Another landmark that washed away that day was "Red Mahn's Daytona Beach Resort." It was located on the far eastern end of Lineville Rd. and Bayshore Dr. The water was extremely high that year to make matters worse.

When the water was high, all the resorts along the Bay would experience water problems. And there were several: besides Mahn's, there was Peters Resort, just down the road from Mahn's; Harbor Lights, Listle's, and Jenk's Channel. Often you'd have to wade through water to get to your favorite bar stool, a minor inconvenience.

Red Mahn once ran for sheriff of Brown County. His platform was (tongue in cheek): He could only steal half as much as his political opponents – he only had one hand.

The information on the Lighthouse history was taken from an interview conducted by the "Wisconsin Coastal Management Program" with Mr. Ralph Weber, son of Andrew Weber, Lighthouse Keeper.

Blessing of The Fleet

Each year before the fishing season started, the local clergy would perform the Rite of "Blessing of the Fleet." Shown here is Father Smith, of St. Paul's Episcopal Parish, blessing Everett Marks' boat, the "Betty Lou." It must have worked, to my knowledge there is no record of anyone losing his or her life due to a commercial fishing boating accident out of Suamico on the Bay of Green Bay.

Everett Marks, Altar Boy Ronnie Beasley, and Father Smith.

But when we weren't being little angels, we had the Game Wardens to keep us in line. It was their job to see that we were obeying the laws and for the most part we did. But every now and then you'd want to have a perch fry for friends and neighbors and that meant smuggling some slightly undersized perch past the Wardens. Commercial fishermen were only allowed to have legal fish on board. So sometimes they'd hide them beneath the floorboards or wrapped up in clothing or stuffed in boots. If they got ear that the Warden was waiting on the dock, they'd drop them off up the river and picked them up later. It was a game of cat and mouse. Most everybody played it and often the "cat" won.

A pleasure boat pulled up to Don and Ray Tuttle's boat when they were fishing on Lake Winnebago. He wanted to buy some illegal game fish. They had a special permit from the DNR to harvest only sheephead and nothing else. It just so happened a Game Warden was accompanying them this day, checking on some of the protected species that wound up in their nets. The lake is famous for it's sturgeon and walleye. Those fish were measured, weighed, tagged and returned to the water. The guy in the boat wanted some walleye. Stub explained to him the trouble he could get into if he were to do that and get caught; he could be fined, lose his license and his fishing rig. The guy laughed and said there wasn't a warden within 50 miles of here. The warden reached in his pocket and pulled out his badge. The guy laughed again and said you could buy a badge like that anywhere. But as he looked closer and began to realize it was for real, the smile, and most of the color left his face. He pushed off and high tailed it.

Personal Stories and Biographies

There's more to the story of Suamico Fishing then I can tell. I was only there for a brief time. It's too bad it wasn't chronicled from the beginning. It started long before I got involved, but I can just relate what I remember and what others have shared with me. Most of the real stories have passed away. But before it all goes and nobody knows what went on here at one time, I felt the need to preserve the memories of those people who lived it and so future Suamico residents will know who their forefathers were and what life was all about here at one time. It was a time when livelihoods revolved around bounties of the Bay. A time when everyone knew each other and most were related in some way. We were like a family, the family of "Suamico, the Fishing Village."

The Author Dan Williquette

That's me, at age 15, sitting on the bow of the "Betty Lou" outside Long Tail Point with the old stone light in the background.

We had some visitors on the boat that day that wanted to see what lifting nets was all about. They took the pictures and were thoughtful enough to make copies for us. It's a fond remembrance. I only wish I'd have had enough sense to take more photos. My thanks go out to all the people who did though and were generous enough to share them with me to make this book more homespun.

The following is a collection of short biographies of families who fished out of Suamico in the mid 1950s. Some were written entirely by family members, others by myself with family member's input and some solely through my memory.

Remember, this whole book is based on memories and not cold hard facts. Some of my recollections may conflict with another person's version or be totally wrong all together. But as Red Skelton used to say; "I calls um the way I sees um." Hopefully it accomplishes what I set out to do; preserve a period of Suamico's past and the memories of the people who lived it.

The Benders

The boat being launched is the "Andy." When Ralph Bender returned from Korea in 1953 he bought the last boat George Lince built. Ralph and his brother John fished together and John named the boat after a big black dog.

The launch sight is the mouth of the Suamico River. On the north side of the river was a public piece of land where people swam and launched boats. The building on the other side of the river is their uncle's cottage. The "Grey Goose," owned by George Lince, is attempting to pull the boat from the trailer. Dick Devroy recalls the event and tells of the trailer getting stuck and lots of things going wrong.

Ralph tells of buying a piece of stainless steel for the roof, and before it was fastened it blew off on a test run at the mouth of the river. Ralph dove in and retrieved it but John lost a Zippo lighter in the process. John asked Ralph if he'd look for his lighter but Ralph said he had enough to do recovering the expensive top.

The "Andy" being launched. Identifiable in the picture are Ralph Bender standing in rear and Francis Euclide next to cabin.

Johnny was the go-to guy. If you needed the impossible done, you went to John. We always called him "Bender on a bender." He had a little trouble with the Oconto County authorities once.

Johnny was the guy if your boat was sunk, or your truck went through the ice or if you had trouble with your flat head Ford V-8, he'd be there. He'd dive in if your anchor line was wrapped around your propeller. When it was time to pull the boats out for winter he was there with his tractor.

We were sitting at the White Front Tavern one night, sharing a cold one, and John said he had a need for a mold to make lead sinkers for gill nets. I worked in a paper mill and had a good friend in the machine shop. We designed it right there on a napkin. It turned out 90% perfect, just needing some minor adjustments. I wish I knew what became of it; I'd like to have it for my fishing memorabilia.

(L to R) Dan Laubenstein, myself and Dan's son Jim. A Sunday afternoon crab cook at Bender's Landing.

I miss the times when he'd call me up and tell me grab my guitar and come down for some crabs. He had a marina at the time called Bender's Landing. A good friend of mine, Dan Laubenstein parked his boat there and he played guitar and accordion, and along with Johnny's permanent guest, Jeff Hermes who played guitar, we'd sit around, make music, eat crabs, drink a beer and have one heck of a time.

The "Cataract," John's gill net boat.

At the time of John's death the "Cataract" was the longest continually licensed commercial fishing boat on the Great Lakes.

John was bragging up the "Cataract" to Donald Tuttle one time, telling him what a great boat it was. Don joked, "do you know what a Cataract is? It's an "eyesore."

The Benedyks

The minnow business had a home in Suamico too. Stanley and Mabel Benedyk had a major operation consisting of three holding ponds, 50 feet wide x 300 feet long x 15 feet deep. Francis Peters, of Peters Trucking, dug the ponds.

The ponds were located east of the tracks adjacent to Vickery Village. They were feed by springs and also by a flowing well on the property. The well flowed year round and also supplied the three holding tanks where minnows were kept for drive up business. Vacationers would stop on their way up north to stock up on minnows and suckers for fishing inland lakes. They'd come with milk cans with aerators to keep the fish fresh for a week. People fishing the bay would also get their bait there.

Stanley Benedyk and an unknown helper seining one of the two ponds next to the train trestle with Mabel and friends looking on.

The minnows consisted of chubs, shiners, golden shiners, cardinals and suckers. The suckers were used for musky fishing and were much larger than the minnows. The seining and trapping was done mostly in Kewaunee, Stiles, Suamico and Sobieski.

Leona Mealy assisted in the operation. She had her own minnow truck and besides catching minnows she transported the fish to bait shops all over North Eastern Wisconsin. She hired local teenagers to help as did most of the fishermen of that era, Dale Leanna, Glen Valentine and Ronnie Cornell to name a few. Ronnie went on to marry Stanley and Mabel's daughter Maryanne. They had another daughter, Patsy and a son Donnie. Mabel was also one of those people from fishing families who worked in the fish houses. When Stanley left the business, Frank Kalfas took over and ran it until they ceased operation.

Mabel and Stanley seining one of the minnow holding tanks. The minnows were transferred to the tank trucks in the background. A very young Donnie and his grandpa are helping too.

Donnie supplied the material for this piece of Suamico history but he almost wasn't around to tell it. His grandpa spotted him bobbing up and down in one of the ponds when he was six years old and his father jumped in and pulled him out.

He cheated death another time while riding with Leona Mealy and Dale Leanna. They were riding in Leona's minnow truck when they were involved in a head on collision. Donnie lost most of his teeth when he hit the dash. The other driver lost his life.

The Borowitz Family

"When we first moved to Suamico, 70 years ago, my father and brothers were always interested in fishing. Shortly after they got a great idea to build a fishing boat. They started with wood, stove, glue and nails and manpower. It took all summer and many men (all neighbors) to complete it. In the fall it was done, launched in the Suamico River with many people watching. The boat was named the "Dorothy B." for the many meals mother made.

The boat was on the waters of Green Bay for many years and caught many fish. Whenever we could, Mildred and I went along for the ride and saw many things like great storms (scared), lots of fish, and seagulls.

Very hard work! It went on for 30 years. Finally sold it and retired."

Marion (Borowitz) Vandersteen

The boat was built next to their barn under the direction of George Lince. The white oak used was harvested on the east shore in Kewaunee County. The boat was later owned by George Tuttle Jr. and rename the Jerry for his son. It wasn't unusual for a boat to have several owners over a lifetime and several name changes.

Marion's husband Norbert decided he'd give fishing a try when he returned from the service. He said he worked with George Lince for three days and that was long enough to find out that the fishing life wasn't for him.

Anton and Ted Borowitz aboard the "Dorothy B." Brother Joe was also part of the crew.

Foot Note;
Anton and Dorothy Borowitz, Sons, Ted, Vic, Joe. Daughters, Selma, Jenny, and twins Mildred and Marion. Lived near Tremble tracks and school. Route 4, which is now School Lane.

The Brunettes

As far back as I can remember, the Brunette families have always been fishermen and farmers. Oliver and Dempsey (brothers) were both commercial fishermen. Oliver's sons, Joseph, Walter, Norbert and Melvin followed in their father's footsteps. Joseph lived and fished out of Marinette most of his life. Walter and Melvin fished out of Suamico, Pensaukee and Oconto. They became partners and became "Brunette Brothers Fisheries". They sold their catches to various fish houses; Lyle Valentine, George Tuttle, Drzewiecki's, Mike Buckarma and Seaway Foods. Norbert helped Walter and Melvin part time as he worked for the railroad.

The Brunette women and children were also involved in the fishing industry. The women, Mary (Walter's wife), Margie (Melvin's wife) and Doris (Norbert's wife) filleted and dressed fish for the above mentioned fish houses. They cleaned all kinds of fish, mainly perch, smelt, carp, northern and walleye. Later Walter and Melvin started their own fish house and the wives worked for them too. The children worked with their Dads on the Bay.

Walter owned two boats, the "Tateter Z" and the "Jennie Lou." Melvin built the "Margie B."

Later when Daniel, Dick, Larry and Paul, (Walter and Melvin's sons) became older, they also fished commercially. Daniel purchased the "Jennie Lou" from Walter and turned it into a trawler and fished for alewives on Lake Michigan. He and Larry fished out of Sheboygan and sold their fish to Swaer's in Pensaukee. They also started their own fish company in Oconto. Paul and his son continue to fish part time.

Oliver retired and mended and patched nets for his son's Melvin and Walter until his passing. Dempsey retired and fabricated nets for Melvin and Walter even while confined to a nursing home where he resided until his death.

Joseph and Walter are deceased. Melvin and Norbert are retired.

Nancy (Brunette) Krause, Walter's daughter.

Standing in front of Walter Brunette's boat, the "Tateter Z.," are John Knoll (L), Nancy Brunette, Annie Knoll and Walter Brunette. The Knolls are Walter's wife, Mary's (Knoll) aunt and uncle. The photo was taken on the original Oliver Brunette homestead in Suamico, later owned by Walter and Mary.

The Devroys

The Devroys were very prominent in Suamico commercial fishing. Art, and his two sons Bill and Dick and Art's nephews Harold (Dulie), and Russell (Bucky) at times fished with each other and with different partners. Art's long time partner was Ed Roderick. Bucky's was Clarence DeBauche and later, Tom Peters. Dulie fished with Art and Bucky and Art's son Bill, and for a brief time with Norm Williquette.

Dick recalls his Dad hooking horses to a gill net shanty and pulling it out to the bay to lift drop nets through the ice. The reins went through holes in the shanty to the horse's harness. On the way he'd pick up Dulie and Bucky.

They had a full cabin boat called the "Mutt and Jeff." Art was the first fisherman to shorten the cabin to facilitate lifting drop nets and eliminating the need to pull a rowboat.

Art's sons Bill and Dick continued the tradition into the 50s and 60s. Bill fished Norm Williquette's rig during the 50s and his own into the 90s. Dick bought the "Arbutus" from Bucky and fished with various helpers, mostly local kids, Darold Rogers, Lyle Kirby, Mike Williquette and Gary Rhodes to name a few.

Dick and one of his helpers were heading out on the "Arbutus" to make a lift, riding a "dead sea." A dead sea is the remnants of a nor'easter. There's no wind or white caps just the rolling waves left over from the storm. It was fun at first, the kid was saying, "wee – wee", it was like a ride on a roller coaster. When he started turning green, Dick turned the boat around and brought him home. He was sick for three days. He was lucky he was with Dick; most guys would have kept him out there till they were done lifting. (I won't say who the kid was cause I don't want to embarrass Darold.)

Dick was a Dr. Doolittle for kids with too much time and too little to do. He'd find something to keep them from getting in trouble. I credit him with keeping a lot of kids from going bad. He and his wife Teresa never had kids of their own but they were a positive influence to many.

Dick was very innovative. His quest at one time was to build an airboat. He called it his "flying machine." He mounted a car engine on a flat bottom scow with an airplane propeller. His first attempt turned into disaster. He had it pulled up into a sluiceway between the river and the carp pond behind our house. When he started it up the propeller hit the concrete wall and shattered the ends. My Dad was there and upon seeing that he named the boat the "Squalus," after a US Navy submarine that had sunk in 1939. He repaired the prop and the next time he tried it, I was with him. We started out down the river but the faster we went the lower the bow sucked down until it almost went under. I was sitting in the bow and kept hollering at him "slow down, slow down, we're going under!" After a few more refinements he achieved success. He was now known as "28 knot Dick", (in his own mind). That's how fast he figured it went.

He took it to Jenk's Channel one day. The back part of the bar faced the Bay and had large windows that opened up to provide ventilation. The channel was just outside the windows. When he left, he opened the throttle and the propeller wash blew everything off the back bar and anything the patrons had sitting on the bar. Irene got a little upset. But it all blew over. (No pun intended.)

Bill was the serious one of the bunch. He tended to the needs of the operation. He made sure the nets were mended and that the small things got done. But he had his moments too. One day he failed to wake me up and Rollie and I sat on the dock waiting for him. When it was obvious he wasn't going to show, we went up to Gokeys to hang out. There we found Bill, in the back room sleeping. He had been out the night before.

Being on the Bay with a hangover is not the place to be. Bill and Dick's brother-in-law, Peg, tried it once. My brother Mike waved an old dead sucker under his nose and he immediately headed for the side where he stayed until they mercifully dropped him off at a lighthouse to let him sleep there until they were done lifting. Suamico boys do have a heart.

Dick Devroy showing off his catch.

Dick's garage doors at his home at Chute Pond. He named the boat on the left, the "Becky" after my daughter Becky Murphy, and the one on the right the "Maggie" after-her daughter and my granddaughter.

The "Arbutus." Jimmy Tuttle had just bought it from Dick. It was sunk in the river. Only the point of the bow and part of the cabin was sticking out as can be seen by the mud line across the bow. Jimmy restored it and used as a pleasure boat.

Bill Devroy all dressed up for the 1961 NFL Championship Game, which the Packers won 37 to 0 over the NY Giants. My little brother, Jim, nine at the time, isn't sure what he is and isn't taking any chances.

The DeWitts

After serving their country during World War I, brothers Joe and Louis DeWitt came home to Suamico and became commercial fishermen.

Their first boat was the "Grace V.," a full cabin boat. These boats were the forerunner to the drop net boat. The full cabin boats needed to tow a row boat to lift drop nets. In 1933 they commissioned Herman and Paul Lade to build a drop net boat, the "Scooter D." They used it until 1948 and then had George Lince built the "Beverly D."

The "Scooter D." with Scooter (Joe Jr.) at the helm.

During the early 40s and 50s in the winter months they made and set pond nets on Little Tail Point and in Pestigo to fish for smelt and herring.

I fished with them until 1949 when the fishing dropped off dramatically. I left and went to work for Leicht's Transfer and Storage. They continued fishing until 1959 when they sold everything to Lyle Valentine. He hired a crew and fished for a couple years and then sold it to Everett and Eugene Marks.

Joe DeWitt Jr. (Scooter)

The "Beverly D."

Joe and Louis took a young Curt Nelson out with them once to groom him as a helper and have a little fun doing so. With each catch there'd be a variety of fish, some you kept, some you threw back. One they didn't keep was the eelpout, a fish similar to a bullhead in that it doesn't have scales and is very slimy and difficult to handle. They told Curt to throw them out but be careful because they were poisonous and to stay away from the head cause a bite could kill him. Curt was determined to throw the fish out but was scared out of his wits as to how to do it. They told him the only safe way was to grab it by the tail. But because the fish is so slippery they knew it would be impossible, but let Curt try until they had their laugh and then clued him in.

Curt Nelson

The Dombrowskis

In the 1930s my grandfather, Walter Dombrowski and my Father Leonard Dombrowski, became partners in the commercial fishing business seining for carp and perch. They made their own nets, tarring them and putting them out in the field to dry.

In 1946 my uncle Charles Dombrowski returned from the service and became partners with my Dad when Grandpa Walter retired.

That winter there was an abundance of herring on the Bay, but deep snow and thick ice made fishing very difficult. A neighbor, Dan Casper Sr. who owned a machine shop in Suamico, fabricated a drill, which they attached to the front bumper of a Model "A" Ford snowmobile. With that they were able to get through the snow and cut through the four feet of ice to set their gill nets.

In 1950 after a long day of fishing on the "Alan" our fishing boat, we were heading for home. Dad and Charlie wanted to rest their eyes so they let me drive. I was to wake them up before we reached the Suamico River. Being 11 years old and thinking, "I can do this". As we approached the river I rounded the first buoy and the steering went out and we ran aground. Needless to say my career as a "Captain" came to an abrupt halt.

Working hand and hand, fishermen needed to have respect for the water, sharp eyes, strong arms, good nets, a sense of humor and lots of faith. Their weathered skin was proof of their hard work, which fed their families. Their interesting tales made memories to share as families and friends gather at the now famous Friday Night Fish Fries of Suamico and surrounding areas.

Allan Dombrowski

The "Alan" parked at Jenk's Channel with a very young Barbara Kozloski, Alan and Alan's Mom Gladys.

In the summer of 1942 Walter and Leonard were searching for some lost fishing equipment south of the Suamico River when they discovered the body of a Coast Guardsman, who had recently fallen from a boat and was presumed drowned. The two men along with Al Drzewiecki transported the body by rowboat to a landing on the Suamico River where Sheriff and Coast Guard officials made positive identification.

The Drzewieckis

The Drzewieckis trace their Suamico roots back to around the turn of the 20th century when Valentine Drzewiecki moved to Suamico from Milwaukee. After being in real estate in Milwaukee he purchased property at the end of Harbor Lights Road along the Bay and took up farming. He started with a small dairy herd and was well known for his strawberries. He took up commercial fishing as a sideline. He had plenty of help from his 11 boys, one girl family. Eventually fishing out-produced farming. They built a fish house on the sight that is now Harbor Lights Resort. When the fish house was no longer needed, one of his sons, Anton, (Tony), turned it into a bar and featured "Point Beer" which he was involved with in Stevens Point. They named the bar after a popular 1940s hit song, "Harbor Lights." Tony and his wife Alzada had the best fish fry around. I can attest to that. Friday nights always began there.

Valentine Drzewiecki on a boat mainly used to haul supplies from Green Bay to the farm. His grandson Val recalls the skeleton of the boat rotting along the shore as a kid.

Besides Tony, there was John, Joe, Phillip, Frank, Valentine (Bill), Bernard (Ben/Chief), Stanley, Aloysius (Butch), Sigmund (Zeke), Walter and daughter Helen.

Butch, Bill, Frank and Joe were the boys involved in fishing. Frank owned two gill net boats, the "Amelia," named after the family's mother and the "Mercury." Joe served as a hand for Frank as did their nephew Horace Drzewiecki.

Frank sitting on the bow of the "Amelia D."

The "Suamico" heading out on opening day with a load of nets.

My father, Bill, fished drop nets with his brother in law Richard Kozloski and also Donald Tuttle and Jack Mealy until I was old enough to partner with him. Eventually Dad devoted full time to the fish house and I started using our neighbor, Ralph Bender's sons, Pat, Kerry and Wayne to work with me as they grew up. My brother-in-law, Jim St. Thomas and Lyle Kirby also fished with me until my son Tom became old enough to join in.

About the time Tom joined in, we abandoned the last wooden boat in the river, the "Suamico," and switched to a diesel powered steel boat. We also switched operations to Oconto. Uncle Zeke refurbished the boat during the transition. I was working full time nights at Armour Meats and Tom was still in high school so fishing was done in the evenings. The fish were plentiful enough that we didn't need to process any to make a living so we sold everything we caught to Anchor and Seaway fish companies.

Soon Armour went out of business so I went into processing our own catch. The perch quota was enough at the time, around 30,000 lbs. But as quotas dropped during the 1980s to as low as 2,500 lbs., Tom went into the marina business and I started buying farm raised perch and doing custom filleting for sports fisherman who were catching tons of perch from Lake Winnibigoshish in Minnesota. Today we process our own catch plus I fish and share the harvest from two private fish farmers which is enough to fill our orders.

Val Drzewiecki

Drzewiecki's fish house, located on the north end of Long Tail Slough, was originally built around 1920. Sometime during the late 1940s or early 50s it was turned into what today is Harbor Lights Resort. (View from the waterside).

Butch Drzewiecki bought the present-day Suamico Fish Co. from Charles Tuttle and later sold it to his brother Bill. Bill's son Val currently owns the business.

The last surviving fish house in Suamico. The Suamico Fish Co., located on the Suamico River.

Francis Euclide

Dad was always an avid fisherman. Grandma Euclide required church attendance Sunday morning, but he invariably borrowed a boat Sunday afternoon and went fishing for herring. Many times he was lucky and Grandpa Euclide smoked and sold these herring for extra money during the depression.

Dad got a commercial fishing job working with George Ladrow in 1935-6. He decided to go into business for himself; that is when the "Roy" was built. Herman Lade built the "Roy" in 1938-9. In the picture I don't see an exhaust pipe poking through the cabin so I suspect this was the initial launching.

He worked with many people, but his longest lasting partner was a character named Jack Mealy. Jack was a brother of George Ladrow's wife Algie. They lived across the street from West Lakeside School. Dad told a story how the "Roy" almost hit a barge in the middle of Green Bay, because Jack fell asleep at the wheel while Dad was washing the floorboards. He told a story how he and Jack decided to cross the Bay going west from Red River into a dangerous wind and thunderstorm. Jack said "nothing would scare him off this frog pond." The incident did scare Jack. Jack always listened to Dad about Lake travel after that. But Jack did not listen to Dad concerning road travel. Eventually Jack was killed on the road by a large truck.

Dad also had many interesting stories about Don Euers, the game warden. Don was in Dad's boat and noticed a rifle in the cabin. Dad said it was to scare away the seagulls. Don declared the rifle illegal and then was promptly "bombed" by a seagull. Don said "I just legalized the rifle."

One day Dad and Don were at George Tuttle's fish house. This fish house was just east of Jim Vickery's store. Don said Dads perch were mostly illegal (too short). Dad said they were legal when he sorted them on the Lake but became shorter as they were brought to the fish house. Of course Don didn't believe this. Dad asked Don to be on his boat the next day. Don was extremely careful when measuring his fish with Dad on the Lake. They brought Don's sorted fish to the fish house and well over half of Don's fish were illegal there.

Dad said at one time 32 boats fished from the Suamico River. Unfortunately the Conservation Department (now the DNR) planted alewives in hope of reviving the salmon sport fishing and ruined the herring production. Dad then switched to perch fishing, which was lost in the late 1940s. Some retired from fishing; Dad went to work for Northwest Engineering which also was lost shortly after Dad retired in 1980.

Roy Euclide

Launching the "Roy."

Emily, Roy and Francis Euclide aboard their brand new drop net boat, the "Roy."

The Hermes

Brothers Leo and Norb Hermes began seining carp with their father Arnold in the 1920s, up until the time of his retirement, and then went on to fish with their sons. Leo formed "Hermes' Fisheries" with his 8 sons; Chuck, Lee, Mick, Laurence, Robert, Dan, Steve and Don.

Norb had 2 sons, Tom and Jim. Jim went on to fish with his three sons, Tom, Bill and Pat, and together they started "Hermes Bros. Fisheries."

Leo Hermes piloting the "Rambler" on the Fox River. In early years they fished out of Green Bay. In 1960 he moved the operation to Duck Creek.

Lee Hermes showing some of the big ones that didn't get away.

Of Leo's sons, only Lee continued the family tradition with his sons Scott, Mark, Bill, Matt and Pete, all of whom are now stockbrokers and no longer in the fishing business.

Norb's son Jim still fishes with his sons, but now they fish for perch.

In the spring, seining was done inside Long Tail and in Peaks Lake. During the summer months the hot spots for carp included an area called the "Bunch" off Grassy Island and the "Frying Pan" located roughly between Pt. Sable and Long Tail Point east of the channel.

The 2,000 ft. seine was pulled with a power wench attached to an anchored rowboat. The fish would end up in the bag portion of the seine and transferred to live racks. A typical live rack held 3,000 lbs. A typical catch would fill 10 to 12 live racks.

On one not so typical day they were heading home around dark with live racks in tow when they noticed Danny was not in the boat! One of the boys thought he'd heard a splash so they circled back and there was Danny floating in the middle of the Bay with only a heavy sheepskin jacket to keep him afloat.

Rounding up the carp for shipping.

Once home, the fish were transferred to their carp pond off the Duck Creek River and held until shipping. Two weeks prior to the Jewish Holidays a broker would determine the fat content of the fish and have them double up on the feed corn. They were then shipped live to the east coast in railroad cars. A car could hold 20,000 lbs. A person would ride along to make sure the fish arrived alive.

Other carp would be shipped to a processor who would make Gefilte Fish, a traditional Jewish dish. The process called for 80% carp and 20% pike or White Fish.

When they weren't destined for Jewish tables the carp would be shipped live to Southern Illinois, Missouri or N. & S. Carolina where they were used to stock "live ponds." Live ponds charge people a fee to catch the fish using hook and line.

Carp ceased being the Hermes' main livelihood in the mid 70s when the FDA determined that the PCB levels in the carp were too high for human consumption and the operation closed down.

Lee and Glory Hermes

Lyle Kirby

Lyle Kirby, Clarence Malchows grandson, got his introduction to fishing by living across the street from Dick and Bill Devroy. It was hard work and very little pay, but they threw in a lot of life's lessons for free. It was toward the end of the era when a man could support his family strictly by fishing so most opted for full time jobs and part time fishing.

After the OJT with Dick and Bill, he went to work for Bill Dickinson, who had taken a job in a paper mill, so he fished mainly with Bill's helper, Reggie Netoles. He fished briefly with Bucky Devroy and Johnny Bender and then for about 4 years with Val Drzewiecki. That ended with the onset of the alewives invasion in late 60s. Lyle said there were dead alewives every square foot of the Bay. Both he and Val stopped fishing and got regular jobs. After that he fished gill nets part time in the winter months and with Freddy Valentine in the summer.

All the time he was fishing he was also a "Farrier," a person who tends to horses hooves, shoeing and trimming etc. It's a trade he still practices today along with his huge veal operation and rearing perch.

Lyle (L) and Donald Tuttle seining one of the ponds while Lyle's mother Audrey looks on.

On his farm he has three large ponds. When he first got into the fish rearing, he raised the perch to eating size. With the help of other fishermen, Val Drzewiecki, Johnny Bender and Donald Tuttle, he'd seine the ponds to harvest the fish. He later found it to be more profitable to raise fingerlings to sell to private pond owners for stocking their own ponds.

Lyle and his wife Diane, and his son Nick (L), pulling in the seine

The results of the pull.

Today Lyle and Dianne take foster children into their home where the kids learn respect, discipline and responsibility, through their example in a stable family environment.

The Kozloskis

Tony Kozloski made crabbing a family affair. His children, from Art the oldest, to Kenny the youngest, and all those in between, Pete, Barb, Geri and Billy, knew three of them would be spending their summer vacation working with their dad on the Bay lifting crab boxes. From the time school let out in June till it started again in September they'd be on the Bay lifting anywhere from 500 to a 1000 boxes a day, every day. The boxes were set inside Little Tail slough, within a square mile, and there were 5 gangs. Besides selling locally, the crabs were shipped live to Nebraska, Chicago and New York.

Tony Kozloski

Barb recalls an incident where she was steering the boat and made a sudden move causing her Dad to lose his balance and fall overboard. By the time they fished him out, he'd lost his wallet with his paycheck from a local paper mill where he worked full time. A few days later Walter and Micky Brunette found the wallet in a net they were lifting off Little Tail point. Every thing was still in it.

Barb (Mrs. Ed White) is also an accomplished taxidermist. She has hundreds of stuffed creatures on display in her home. It's a hobby she's enjoyed for some 20 years. Joining her and learning trade is Lynda Van Boxtel.

Barb and Lynda and some of their work.

Barb and "Melvin The Moose."

Her favorite piece is "Melvin The Moose." It was given to her already mounted. After much head scratching and a bottle of vodka, she and some of her friends figured out a way to maneuver it into her bedroom where it hangs proudly on the wall, taking up half the room.

Barb (Kozloski) White

George Ladrow

George Ladrow. Everybody called him unk. And he was uncle to lot of people in Suamico. He was one of the more successful fishermen, making it big when the herring were plentiful. I don't know how rich he was but he always drove nice cars.

He always seemed like the measuring stick as to who was successful and who wasn't in the fishing community. He was wealthy enough to lend my father the money to buy our house and the money to renovate it. The first thing my mother did, after my father died and she received his insurance check, was to pay off George.

When I think about it, there was only one Ladrow in Suamico. All the other families had multiple fishermen to their namesakes. He was a rarity.

He and his wife Algie had no kids of their own but they took in Reggie Netoles. Reggie was slightly challenged but not enough to keep him out of WWII. After the war he fished with George. Before that George fished with Dempsy Brunette and many helpers over the years.

Picture taken probably in the 20s. I've been unable to identify any of the people on board.

His first boat, the "Algie L.," disappeared somewhere off Little Tail Point. It was late in the year and they were returning to Suamico from fishing in Oconto when the ice got too thick to make it into the river. They left the boat near the point of Little Tail and walked home on the ice. When they returned the next day to retrieve it, the "Algie L." was gone. They assumed the ice moved and knocked a hole in it and it sank.

One early spring the Coast Guard announced it was sending an icebreaker through the south end of the Bay before the fishermen had a chance to pull their nets. George contacted the local C.G. and asked them if they could put it off. The fishing was good and there was still adequate ice to work their nets. They said they didn't have the authority to change their orders and that he'd have to contact Washington. He did and they got their delay. Some clout!

His next boat, the "Mary Lou" was built by Herman Lade.

George Ladrow (L), Reggie Netoles and Mickey Brunette.

Bill Dickinson

When George Ladrow retired he turned the rig over to his nephew Bill Dickinson. Bill hired Gerry Shallow and myself to fish with him. I was paid $50 dollars a week. I had a 1950 "Olds 98" at the time and every Friday there'd be a check on my steering wheel. Billy's mom, Sis, took care of the books and made sure we got paid.

When the fishing went bad on Green Bay, Bill went into the heating and air conditioning business with a sideline of selling Moto Ski snowmobiles.

But fishing was in his blood and the lure of big perch catches on Lake Michigan soon had him purchasing a new fishing rig and moving his operation to Michigan City Indiana. Tragedy struck there when his boat sunk in a storm and one of his crewmembers lost his life.

Bill's Lake Michigan drop net boat.

Bill (L) and a crewmember pulling in a pot full of perch.

He moved again, this time to Florida where he fished and crabbed in the Gulf of Mexico until a stroke in 1995 prematurely ended his career. He hired a crew to fish his rig with the option to buy. Once sold, he retired and moved back to Suamico.

The Malchows

The Malchow's journey to Amercia originated in Malchow, Germany in 1866. Great Grandfather Ludwig Malchow came to this country with his family, his wife and twelve children, not to a new frontier, but to a new life on the waters to continue his occupation as a fisherman. After arriving and settling in Chicago, Illinois they decided to venture north to Oshkosh, Wisconsin in 1872. Ludwig was known as one of the pioneer fishermen of that city. He was one of the first fishermen to ship fish from Oshkosh to outside markets. He was the first to smoke fish. He taught his sons the fishing trade and sent them out to spread their knowledge. He died in 1902 at the age of 75 years.

His family wasn't above breaking the law *or* enforcing it. An excerpt from the April 1885 "Appleton Post Crescent" reads: "Fish Wardens issue warning of law forbidding the catching of fish with nets in the Fox River and Lake Winnebago." It is well known the law is violated with impunity every year and vast quantities of fish are netted and shipped to Chicago. An excerpt from the August 1887 "Post Crescent" reads: "Game Warden Malchow of Oshkosh saw his father fishing with a net and promptly arrested him." He was the first of three generations of Malchow Wardens that include Gordon and his son Vernon.

The first son born to Ludwig and Augusta Malchow was our grandfather, William. He was born in 1854 in Germany and was twelve years old when he came to America. On Thanksgiving Day in 1877 he married Johanna Marheine and began raising a family.

"Malchow Heritage Farm."

In 1885 he packed up his family and moved to Green Bay where he worked as a foreman in the fishing industry along with teaching other fisherman how to smoke and successfully freeze fish.

In 1886 opportunity knocked and he answered the call by settling in Suamico, WI. He purchased land on the west shore of Green Bay with one mile of shoreline. The property became known as "Oak Ridge Dairy Farm." In 1982, at a family reunion, it was changed to "The Malchow Heritage Farm." It was there they raised their eleven children. Six of the seven boys followed their father's love for the land and water.

William Jr. moved back to Oshkosh where he became one of the youngest licensed boat captains to sail the waterways.

Walter moved to Racine, Wisconsin, where he established a fish market where he wholesaled fish purchased from Suamico fishermen including his brothers Clarence, Charlie and Gordon.

Arthur left Suamico and went to work in the mines in Chrystal Falls, Michigan. His heart and instincts brought him back to Appleton, Wisconsin where he became a Master Lock Tender on the Fox River Lock System.

Charlie, Clarence and Gordon Malchow and a lot of dead ducks.

Charlie was the most rugged of the boys. He stayed on the farm and in 1917 joined the service during WWI. In one of his letters he wrote to Grandma Malchow from France; he told her to tell Art and Gordon to forget the idea of farming and take his seine and fish more. He even sent a pattern for making nets. Other than his mother, his greatest concern was for Frank Mutz, a neighbor who was stationed in the same war zone. Another concern he had was for his Winchester shotgun and wanted to make sure Gordon was taking good care of it. He was very proud of the fact that he, Clarence and Gordon were the best shots around. He proudly fought and almost gave his life for our country, his new country. Upon returning from the war he stayed living on the farm. He continued his commercial fishing, trapping, frogging, smoking fish and making nets and crab boxes. In later years he went to work for Selmer Construction Co. until his retirement. Charlie and his wife Pearl had no children of their own but treated Gordon's children as theirs.

The "Edward C." Malchow's fishing boat. It was named after a little boy who loved to visit the farm and see his "Uncle Bill." The little boy grew up to become a highly decorated WWII hero in the U.S. Army, Brigadier General Edward L. Carmichael.

Clarence Malchow. (As told by the author.) Clarence and Martha lived across the street from us. Besides raising their own family they also raised two of their grandchildren, Lyle Kirby and Susan Malchow. Next to his family, Clarence's greatest love was for his Chesapeake hunting dogs. A fervent hunter and dog trainer, he always said, "I wouldn't eat a duck unless it was shot right through the eye." He might have been stretching it a little. And he also called my brother Mike his "little pismire." A pismire is an ant, but before I knew better I thought it was something worse. In the winter, during the depression and WWII, Martha would string gill nets for fishermen to earn extra money. She also had a loom and made throw rugs out of rags to sell. In the summer months she worked for Stokely Canning Co. in Duck Creek. She always had the smell of sweet pickle about her. Clarence retired from commercial fishing and became Lock Tender for the De Pere, Wisconsin Lock of the Fox River Lock System. Trustees from the Reformatory (Green Bay Correctional Institution), would provide the manual labor and Clarence would oversee the operation.

Lyle Kirby asked me to include this in his grandfather's memoir: My father, Norm Williquette would visit Clarence every Sunday morning after church. Clarence had suffered a stroke and was confined to a wheel chair. After the visit my dad would say to Clarence, "I'll say goodbye now cause you may not be here next week." As fate would have it my father passed away before Clarence did. Clarence took it very hard and passed away the following year. (End of author's contribution.)

Gordon was the youngest of the eleven children. As a youth he worked as an extra hand on the fishing boat but he was destined to continue the family farming operation. Being a dairy farm, there was milk to be hauled. And living so far from the creamery he decided to start a milk route and pick up all the farmer's milk on the way to Giese's Creamery in downtown Suamico.

He expanded his route to include deliveries to Fairmont, Aquity and Dietsch Dairy in Green Bay. Another stop was the Northwest Co-op, which also had a grocery store. The ladies on his milk route would give him a shopping list in the morning which he'd fill while the milk was being unloaded and then deliver it to them, along with their empty milk cans, in the evening. When he wasn't hauling milk he'd haul potatoes and cabbage out west. He obtained one of the first Contract Carrier Licenses issued in the State of Wisconsin, license no. 1329, which is still owned by the family.

Tragedy entered his life when his first wife passed away leaving him alone to care for their sixteen month baby boy. Luckily his travels took him to Appleton, Wisconsin where he met Ethel Terrio, the person who would become his soul mate, life long partner and companion, but most important, mother to his baby boy.

The Giese Creamery located across from Vickery's store.

Somehow Gordon found time to manage the best "Little Six Baseball League" team in 1935. The "Suamico Tigers" was made up of all guys from Suamico.

Front Row; (L-R) Louie Lewinski, Ed Malchow (Gordon's son), and John Scepanski.

Cent. Row; Leonard Dombrowski, Carl Hess, Horace Drzewiecki, Peter Kalfas, Walter Brunette and Tony Kalfas.

Back Row; Gordon Malchow, Nelson Hansen, Elmer Hansen, Alex Hock, Art Fiest, George Roderick and Arnold Anderson.

Gordon continued the family's hunting tradition by teaching his wife and children the art of living off the land. At mealtime, the main course was usually some kind of game, be it from land, air or sea.

Being an avid hunter he soon discovered he had a knack for training hunting dogs. So good that his Black Labrador, "Dixie Dinah Of Landover," earned his "Oak Ridge Kennels" a Grand National Champion in 1944. He also boarded, trained and showed dogs at shows and trials.

Gordon with his Grand Champion Chesapeake " Cocoa Belle."

By now his family had increased in numbers and living so far from school meant driving them to school everyday as did everyone else living in the Long Tail Beach area. The purchase of a carryall van to fulfill that need was the first step toward the formation of the "Malchow Bus Service" which serviced the Suamico and surrounding area for 51 years.

(A note from the author.) I was privileged to ride Malchow buses for several years, back and forth to Pulaski High School. I don't ever recall an incident where the driver had to stop the bus to reprimand anyone. The drivers showed the kids respect and in return they earned the respect they deserved. (End of note.)

In 1962 the company was recognized by the National Safety Council for transporting children in the Suamico School District for 22 years without an accident.

. (Lto R) Vernon Malchow, Gordon Malchow, Jim Vandermoss, Joyce Malchow, Mabel Benedyk, and Edmond Malchow, in front of their busses.

Gordon was very involved with Boy Scout Troop 33. He served on the council committee, was an Explorer Scout advisor and taught the Marksmanship Merit Badge. He was also an instructor for the Explorer Scout's N.R.A. Rifle team.

Presently in the year 2007, the original home built by William Malchow in 1886 is occupied by Gordon's widow Ethel and his daughter and son-in-law Eileen and Gordie Olson. Others whose homes are on the property are Gordon's son Vernon and his wife Claire and Vernon's son Bert and his wife Debbie. In 1987 the farm was recognized as a "Century Farm" for being owned by the same family for a hundred years.

Suamico was always very important to Gordon and his family. His wishes are, "The Malchow Heritage Farm" be kept for the restoration and preservation of wild life.

Eileen (Malchow) Olson

The Maricque Legacy

Two became three….Then back down to two.

Legend has it, fishing in the Maricque family started about six generations ago. Unfortunately for Mark and Nick Maricque, the father and son commercial fishing members of the family today, names and specifics weren't passed down over the years.

Mark Maricque just getting started in his lifelong profession.

Mark's great uncle Frank (L) and his father Bob.

Mark became the third member on the family's unnamed drop net boat, along with his father William Robert (Bob), and Bob's Uncle Frank. Mark's first memories of his fishing past are of a gift given him, when he was five, from his father and his Grandpa Vander Geeten (Papa); it was an ash shovel with a seven and a half inch measure attached to the handle, patterned after the shovels fishermen use to measure fish, so he could learn to sort fish when his time came. His lot in life was cast.

Their "unnamed" boat docked on the East River in Green Bay.

During this time period, their boat was docked in Green Bay, Wisconsin, on the East River, where Frank and the late Elmer and Max owned a fish house. With Frank's retirement, the crew was down to two members, Bob and his son Mark, until Bob's brother Jim joined them.

In 1984 the boat was moved out of Green Bay to the Suamico River where it docked at numerous places, including Bill Dickinson's, the Mark's, Tom Peters' and Red Belongia's.

In 1985, Jim removed himself from the crew to work on the other side of the family's fish business, the fish house and Maricque's Bar on University Avenue in Green Bay. Renowned for their Wednesday and Friday fish fries, family members still wait tables and fry the fish that Mark and Nick catch.

The fishing then switched from a three-man drop net operation to a two-man gill netting crew.

In 1998, a new family member entered the picture; Mark's 15-year-old son Nick joined his father and grandfather as the third member aboard the "Frank J.," the family's gill net boat named after Bob's Uncle Frank.

In 2000, "Maricque and Maricque Fisheries" purchased property on the Suamico River where the boat is parked today.

The "Frank J." at it's berth on the Suamico River.

With the passing of Bob Maricque in May of 2002 the "Frank J." was back down to a two-man crew. It stayed that way for a number of years until the Department of Natural Resources cut the fishing quota by 90 %. With far less fish to catch, Suamico fishermen were left with lot of extra time on their hands. Mark and Nick extended an invitation to Tom Peters to join them on the "Frank J." The two became three again.

When it was time for Nick to return to school, be it high school, then collage and now as a teacher, Tom and Mark would continue fishing. In the fall they'd catch perch on the "Frank J." In the spring, once the ice was out of the river, they'd switch to whitefish on Tom's boat, the "Bucky D."

Tom and Mark share a story that happened in the spring of 2007 that neither had ever witnessed before. While lifting their nets out on the Bay, they saw a seagull *walking* along side their boat! Actually a thin layer of ice had developed while they were working. Although they say it was kind of scary, the smiles on their faces suggest a fonder memory.

With the onset of the 2007 season, the Maricques again lost their third crewmember. On July 2, 2007 Tom Peters passed away unexpectedly, leaving father and son to fish alone once more.... back down to two.

Mark and Nick Maricque

The Marks Family

Commercial fishing started for the Marks family in 1921 when Otto Marks got his first Wisconsin fishing license from the State Conservation Commission. The license was for 5-drop nets and cost $6.25. I have all his licenses from 1921 through 1953, he saved them all.

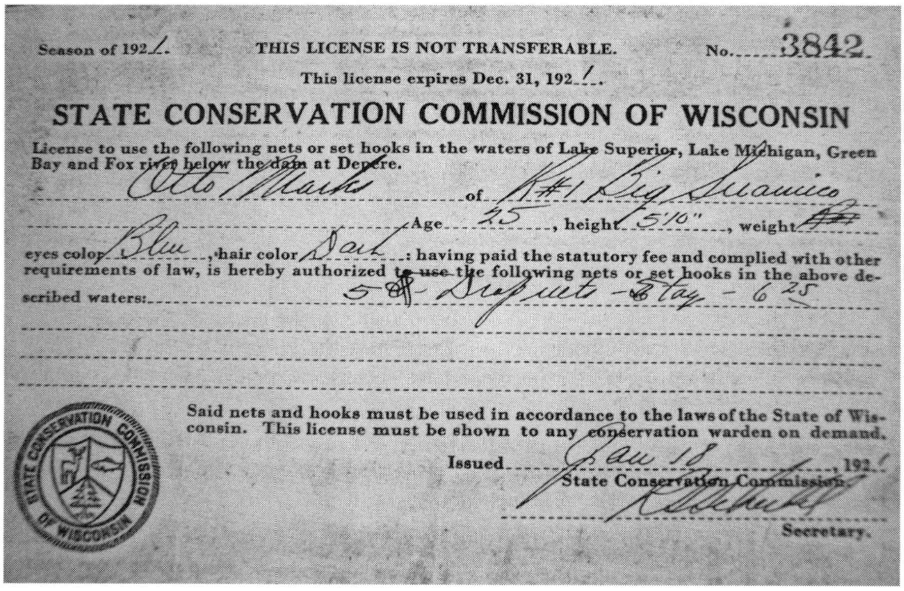

Season of 1921 THIS LICENSE IS NOT TRANSFERABLE. No. 3842

This license expires Dec. 31, 1921.

STATE CONSERVATION COMMISSION OF WISCONSIN

License to use the following nets or set hooks in the waters of Lake Superior, Lake Michigan, Green Bay and Fox river below the dam at Depere.

Otto Marks of R#1 Big Suamico

Age 25, height 5'10", weight [illegible]

eyes color Blue, hair color Dark: having paid the statutory fee and complied with other requirements of law, is hereby authorized to use the following nets or set hooks in the above described waters: 5 [illegible] Drop nets - [illegible] - 6.25

Said nets and hooks must be used in accordance to the laws of the State of Wisconsin. This license must be shown to any conservation warden on demand.

Issued Jan 18, 1921

State Conservation Commission,

[illegible]

Secretary.

He started fishing by himself with hired help from a neighbor Bob Walczyk, who moved to Suamico from Chicago. In 1923 he had help from his brother in law, Floyd Farrell, who he fished with for a few years.

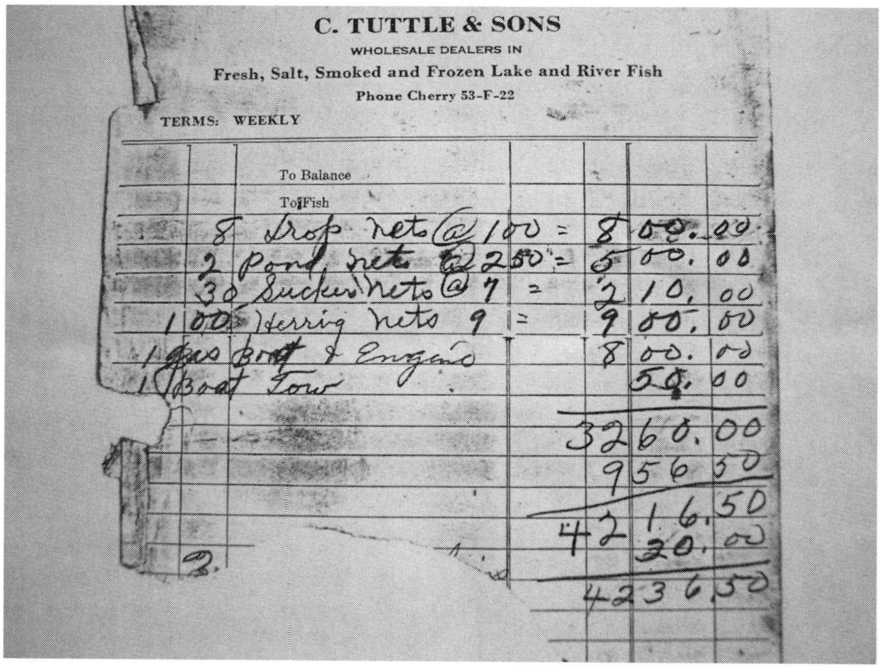

C. TUTTLE & SONS

WHOLESALE DEALERS IN

Fresh, Salt, Smoked and Frozen Lake and River Fish

Phone Cherry 53-F-22

TERMS: WEEKLY

To Balance	
To Fish	
8 Drop nets @ 100 =	800.00
2 Pond nets @ 250 =	500.00
30 Sucker nets @ 7 =	210.00
100 Herring nets 9 =	900.00
1 Gas Boat & Engine	800.00
1 Boat Tow	50.00
	3260.00
	956.50
	4216.50
	20.00
	4236.50

He purchased more equipment from Charles Tuttle. The receipt he saved shows the total transaction. He also made some of his own nets with twine purchased from the Linen Thread Co. and many other suppliers. My brother Gene and I did the same when we fished together.

When Eugene became old enough to handle the hard work involved in fishing he started working with Dad. Around 1940 they purchased a fishing rig from Carl Jenkins. (He and his wife Irene owned Jenks Channel Resort). When I became old enough to pitch in, we had a total of 32 pots. We also ran a small dairy herd and raised cash crops of potatoes and cabbage. Gene and Dad fished until the late 40s. Besides fishing we also put up ice in the wintertime for ourselves, the DeWitt Brothers and George Ladrow.

This all came to a halt when Dad became ill. We held an auction of the farm animals and equipment and sold 80 acres to Vic Peters on County J. Freddy Valentine bought the boat, the "Eugene M." and some of the fishing equipment.

With the fishing operation ended, Gene went to work for Dettman Tire Co. and later for the Green Bay Police Dept.

I started my fishing career with the experience I gained from working with Dad and Gene. I graduated from West High in 1954 and went to work for Lyle Valentine Fisheries.
Things got slow when the fishing production from Lake Erie dropped off, but there were a lot of fish in the south end of Green Bay. Lyle and I decided to buy out his uncle, Jack Mealy's nets. Jack taught me a lot about fishing the north end of the Bay around Peshtigo, Oconto and Marinette. He knew the area well and all the good fishing spots. Jack helped me mend nets in the twine yard we rented from Ben Valentine, the one next to his house and West Lakeside School. We bought John Valentine's boat, the "Betty Lou" and kept it next to DeWitt's across the river from our farm, by Ben Valentines old fish house. That was 1958. Shortly there after we bought DeWitt's rig consisting of 40 pots and their newer George Lince built boat, the "Beverly D." We also used DeWitt's twine yard so we had everything on the south side of the river.

Lifting a pound net with the "Beverly D."

In the partnership, I worked the boat and Lyle worked the fish house. My hired help over time included, Neal Vannieu, Dick Brunette, Butch Deaton and Russell (Bucky) Devroy.

In 1961 my brother Gene and I bought out Lyle and we formed a partnership. Gene was a Green Bay Police Officer at the time and I started working full time at Fort Howard Paper Co. We both worked 2nd shift so that worked out well. In 1990, after 29-1/2 years at the Fort, I quit and went fishing full time.

We fished part time with 2 licenses, #40 and #42. Each license covered 30 pots. By now we had over 60 pots but could only have 60 in the water at any one time. We moved our operation from Dewitt's to our property across the river.

We had help from a lot of young Suamico boys in our operation; Harold La Londe, Curt Nelson, Ken Bradael, Jimmy Lewis, Gene's sons, Dan and Jeff, and my son Joe.

The "Beverly D." was our main boat but we needed a back up so we bought a double-ended boat out of Kewaunee, which came to be known as the "Rubber Chicken." We cut the stern off; put in floorboards and built a cabin on it to make it fit to lift drop nets. It worked, but not very good. We sold it to Allen Hanson who sold it to Bob Nelson. We still see it pass the house on its way to the boat landing with its present owner.

The "Jumpin' Jimmy" and the "Jimmy II".

Gene had the "Jumpin' Jimmy" built. A steel boat with a 671 Detroit diesel engine. It became our main boat because of its speed, spray rail; trim tabs, keel and cooler. An all around better boat. We sold the "Beverly D." to Calvin Williquette.

As the perch quotas increased, we built another steel boat, the "Jimmy II," with an 871 G.M. Detroit diesel. It was wider, faster and could carry a bigger load.

Eugene (L) and Everett aboard the "Jumpin' Jimmy."

When the D.N.R. changed the law requiring gill net fishing only from May 20th until July 1st we were forced into buying and making more gill nets and purchasing a gill net boat. We bought the "Betty Ann" from Paul Spotts. We also purchased his fishing quota as well as Bob Coleman's, Calvin Williquette's and Red Belanger's in the years 1986, 87 and 89. This was a time when the D.N.R. was increasing perch quotas 400,000 lbs. to 475,000 lbs. That was the highest the perch quotas ever got. The quotas dropped shortly after that to 300,000 lbs. in 1994/95 to a low of 20,000 lbs. for the entire fishery. In 2005/6 it came back up to 40,000 lbs. and to 60,000 lbs. in 2007. Not much quota for the amount of fish in the south end of the Bay.

The "Betty Ann."

Our fishing phased out with the onset of the cormorants the D.N.R. assisted in planting on Grassy Island. Their population got way out of control. They claimed they didn't eat perch but were later proven wrong.

Perch is a crop that has to be harvested like any other crop. The fish start eating their young when they get to plentiful which is what's happening today, the spring of 2007. Very little is being done to control the cormorant population, a lot of talk but very little action. As Carl Van Roy, our State Representative says; "It's time to start doing something and quit talking."

Perch fishing has been up and down since my father started fishing in 1921. You cannot control nature. Bad weather in the spawning season, April and May, has a big effect on perch population. The sport fishermen, the general public and the commercial fishermen, who are fast becoming extinct, are all hurt by low quotas when the perch population is at an all time high as it is now. The D.N.R and the politicians need to find a way to speed up the process to raise the quotas and not take two years to do it. They have an emergency rule to lower the quotas if necessary, why not an emergency rule to raise them?
Ask the D.N.R. (The $64,000 dollar question!)

Everett Marks

Jack Mealy and Ben Valentine.

Jack Mealy was one of those in-between guys. As a kid, I never knew who he worked for, where he lived or what his connection to Suamico's fishing was. He never had his own rig but I've learned he'd fished with his brother-in-law Ben Valentine and Francis Euclide, and probably many others in his lifetime. He was always around. You'd see him everywhere driving his Ford pickup with a stub of a cigar in his mouth. His sisters were all married into fishing families; besides his sister Grace being married to Ben, his sister Algie was married to George Ladrow and sister Evie to John Valentine Sr. Another sister, Leona seined minnows for a living.

Besides being a gruff old character, he was known around town as "second gear Jack." He'd often get in his truck and take off and forget to shift out of first gear or second gear. He'd go for miles with the truck engine roaring and be totally oblivious to it. Anyone riding with him would have to remind him it was time to shift gears.

I don't know anyone left to tell about Ben Valentine, so I'll write down what I remember. Ben and his wife Grace lived next to West Lakeside School, just to the east. His twine yard was east of his house and the field to the rear is where we played ball. His tar kettle was near the road. I never saw him wearing anything but knee boots.

He was fishing with Jack Mealy at this time and Joey Vannieu was their helper. They were tarring nets and Joey was driving the twine yard truck. Jack hollered to Joey to "go ahead and back up," which translated meant Jack was giving Joey direction to back the truck up. But Joey, in his mischievous way knew exactly what Jack meant but instead drove the truck ahead and then backed it up. Jack got mad and told Joey, " ya little S.O.B. if ya don't wanna work, go on home!" Ben and Joey got a big kick out that.

Ben had a son-in-law named Earl Berringer who would drive to Suamico every year from his home in Ashland to hunt duck and pheasant with my Dad.

They took me duck hunting with them one morning and the next day Dad woke me up, before the sun was up, to go again. I made up some lame excuse why I didn't want to go. But actually what happened was one day of sitting out in that cold duck blind convinced me duck hunting wasn't for me.

Another time before I was old enough to hunt they stopped home for some lunch after hunting all morning and Dad brought in a bag that he said had two rooster pheasants in it. As they were leaving, I looked in the bag and discovered there were actually three pheasants in the bag. I went running out of the house with the third pheasant to show my Dad. The car screeched to a halt and my Dad came running and rushed me and the bird back into the house. It was a hen and illegal to shoot.

I had a dog-named Rip and had just gotten home from somewhere when I noticed he was across the street. I called to him and at the same time out of the corner of my eye I saw Grace Valentine coming up the road in her 51 Nash on her way back to work at one of the fish houses. There was no stopping the inevitable. I picked him up about 50 yards down the street after impact. He seemed okay to me but my Dad knew better and mercifully put him down. I was crying in my room and never heard the gun go off. I didn't blame Grace, there was nothing she could do to prevent it.

The Mutz Brothers

Brothers Jack (L) and Jake doing what workingmen enjoy - sharing a "cold one."

The Mutz Brothers, John G. (Jack) and Jacob B. (Jake) were long-time Suamico commercial fishermen, and they also caught crabs (crayfish) in the Green Bay waters of Lake Michigan. They came from a large family, but only they and their oldest brother, Frank, were involved in fishing with their father.

When their father passed away in 1943, and their brother Frank in 1948, Jack and Jake continued fishing and crabbing as "The Mutz Brothers" partnership late into their senior, octogenarian years.

Their parents, John Sr. and Martha Mutz, immigrated to the United States from Poland (West Prussia) in the late 1800s. From their ports of entry (Baltimore and New York), they migrated to Milwaukee, Wisconsin, first settling on Jones Island.

The Jones Island settlers were mostly Polish, and those early Polish immigrant fishermen were called "*Kashube*"; members of a Slavic people from the vicinity of Gdansk. They chose Jones Island because the Lake Michigan waters reminded them of the Baltic Sea of their native homeland.

Their stay on Jones Island was rather short-lived, however, because shortly after settling there, the young City of Milwaukee planned to take over Jones Island as a harbor facility and forced most of its inhabitants to seek housing elsewhere. (Eventually the Jones Island community disappeared.) Thus, John and Martha Mutz traveled northward along Lake Michigan to the west shore of Green Bay and settled on acres of waterfront property in Suamico, which became their lifelong residence.

Initially, they engaged in "family farming," raising marketable fruits and vegetables and catching fish for a livelihood. As their family grew, so did the number of helping hands.

As time went by, father and sons became more involved in the fishing business because of the abundance of fish that thrived in the Green Bay waters. In addition to the necessary fishing equipment, two special and practical features for their work were the private fishpond and the icehouse.

In the early 1900s, using a team of horses, a plow, and a "slusher" to move the dirt and wet mud, they manually dug a pond on a parcel of their property approximately an acre in size and eight feet deep. It was principally used for temporarily holding and feeding the fish, especially carp, which were corn fed until ready for market.

The pond was connected to the Bay via a sluiceway, an artificial channel or passage with floodgates for regulating the flow of water.

The fishpond brimming with carp.

When ready for market, the fish were removed from the pond, packed into wooden fish boxes, iced up and taken to the Suamico Railroad Depot for shipment to customers.

Occasionally very special large orders were placed for "live-truck" shipment to Chicago, Baltimore, Philadelphia and New York. Certain clientele ordered only live fish, and it was necessary to have a huge truckload for practical and profitable delivery. Large trucks with cold-water tanks would be driven directly to the pond to pick up their load. Occasionally Jack or Jake would accompany the truck and driver to ensure the fish arrived at the delivery destination fresh and alive. Jake would bring back mementos of his trips to his wife, Anne, and his daughters, Joan and Monica. The girls still have some of them as treasured keepsakes.

Commercial fishing was a year-round full-time occupation, and it was not an easy means of livelihood. Like other fishermen in Suamico and elsewhere, the brothers endured the long hours and exhausting physical labor of their profession. Maintenance and upkeep of the large boats and the many nets needed for their work required a big investment of time and money.

Jack (L) and Jake lifting a drop net through the ice.

Jack "knitting" a new net near the tar kettle.

Jack had a special talent for making fishnets and painstakingly handcrafted many of the nets they used, selecting special "knitting needles" and weights of twine according to the type of fish to be caught. When completed, the nets would be tarred in a huge kettle to treat, strengthen and protect them from the abuse of the waters. It was a monumental project.

In the early years of commercial fishing, there was no such thing as refrigeration or ice machines. The only means to keep something cold and fresh was "Mother Nature's" ice. In order to have a supply of ice available during the warm summer weather, fishermen had an ice-house for storage of ice obtained during the winter from the frozen waters of ponds, the river or the Bay.

The Mutz icehouse was a large wood-frame building with a concrete floor next to the fish house. It was windowless and had an oversized front-door opening for easy access to the pond for making ice.

Every winter a group of men would gather at the Mutz pond to "make ice." It was a big production, requiring great team work to saw cut the ice into large blocks and move them into the icehouse on a conveyor-like slide and carefully layer them into tiers until the icehouse was full. When full, the ice was covered with sawdust, top and sides, to prevent it from melting. The ice was used as needed, block-by-block, to keep the fish and crabs fresh for short-term holding and transportation. The supply would usually last through the summer months and into the fall.

Making ice on the Mutz fishpond for use during the summer fishing season. The car in the background is being used to pull the blocks of ice up the slide and into the ice-house.

Initially ice making was a very labor-intensive necessity every winter. However, with the "advancement of technology," a large circular saw blade replaced a rear wheel on a Model "T" chassis and eliminated much of the manual labor.

Later, with the emergence of refrigeration, the annual ice-making tradition became more of a celebration than a necessity, and the ice crew still looked forward to gathering for the camaraderie and delicious home-cooked meals prepared by Jake's wife, Anne.

Jack (L) and Jake sorting out a day's catch of crabs.

Besides catching fish, the Mutz brothers also caught bay crabs. It was a short, summer-seasonal diversion from fishing that lasted from June until September. Although the season was short, like fishing, crabbing required much work. The crab boxes were made by hand, usually during the winter months when fishing was slower. They were made of wooden slats, so the water could flow through, and the ends had tunnels so the crabs could get in but not out. They were weighted with a bit of concrete so they'd stay on the bay bottom when set. The covers were also slatted and fitted for easy removal of the crabs and baiting. A rope tied to the box and a float completed the unit. The bait was fresh chunks of cut-up carp or suckers.

Over a span of years, the 1950s and 1960s and beyond, the fresh clean waters of Green Bay became polluted and affected both the quality and quantity of the fish and bay crabs. In the good years, prior to the pollution, one could catch several thousand crabs daily; and then the catch dwindled to less than a thousand or less than that. Likewise with the fish, once plentiful, there became a huge decline and even setting more nets didn't seem to increase the numbers of fish caught.

Perhaps there are no specific scientific reasons for the decline, but pollution seems to be the most likely cause. Nature can't survive with deadly, toxic contaminants in the environment.

Yes, commercial fishing was a very common and popular occupation in Suamico for many years. Jack and Jake Mutz were part of that hardy breed of men who lifted those heavy, wet fishnets and endured the hardships of the job well into their senior and octogenarian years. In the late 1970s, Jack's aging body, some broken bones and declining health slowed them down, and eventually their partnership ceased. After that, Jake who was a number of years younger than Jack, continued catching bay crabs and fish on his own with the help of others until the aging process curtailed his endeavors as well.

Jack (L) and Jake at ease in the peaceful surrounding of their Suamico "paradise." Perhaps knowing how to relax had a lot to do with their longevity.

Today, the once thriving fishing industry in Suamico is almost nonexistent. Most of the "old timers" have died and only a few commercial fishermen remain. Instead of large fishing fleets on the Bay, now one mainly sees pleasure boats and sailboats.

And at the Mutz homestead, only antique remnants of a fish house and icehouse, and a pond without a sluiceway (that is overgrown with noxious invasive Phragmites), remain as visible signs and reminders of what was once a rather unique pristine and picturesque setting of commercial fishing in Suamico, on the waters of Green Bay.

Joan and Monica Mutz.

P.S. Hopefully a "cure" will soon be found to control and eliminate the invasive Phragmites which is foreign to our environment and one of the most widespread noxious and aggressive plant species in Wisconsin.

Tom Peters

My parents had a cottage on the Bay in Suamico, when I was growing up, and there is where I developed my love for the water. I started crabbing on the Bay when I was 14. I bought a tunnel bottom boat with a Model "A" engine for lifting crab boxes. At age 16, I bought a few old gill nets. I played around with them until I figured out how they worked, and then decided to get a commercial fishing license.

I started out picking perch out of gill nets for Herb and Arnie Zakowski. And then working with Johnny Bender lifting drop nets. In the evening I fished with Freddie Valentine lifting his nets, when he got home from his day job. For the next two summers I worked for Everett and Eugene Marks.

When I got out of high school I got a job at a meat packing plant. A short time later I was drafted and spent two years in the military. Upon returning from the service I realized the meat packing business wasn't for me. I partnered with Russell (Bucky) Devroy. We had an old drop net boat named the "Jean," named after Bucky's daughter, and 20 old cotton drop nets. Over time we converted to all nylon. We moved up to a steel drop net boat we called the "Cheryl Lee" after my daughter.

We fished pound nets in the spring of the year for smelt for something to do until the perch season opened. Slowly we built the rig up to 60 pots. We bought another steel drop net boat and named it the "Chad Richard" after my son. We tried to fish the gill nets and drop nets from the same boat but that didn't work out. We bought a steel gill net boat called the "Gen." Bucky and Ray (Stub) Tuttle fished the gill nets and I fished the drop nets with half the young boys in Suamico who needed a job and were willing to work. Chris Shallow, in particular, worked with me for several years.

The "Bucky D."

We fished that way until Bucky passed away in 1991. I remodeled the gill net boat and renamed it the "Bucky D." in his memory.

As the quotas dropped, I stopped fishing drop nets and went strictly to gill nets. As the quotas decreased even more, I was forced to fish whitefish to maintain my license. Now the perch are coming back and the quotas are increasing. My new partner is Mark Maricque. We fish whitefish with the "Bucky D." and perch with Mark's boat the "Frank J."

I plan on commercial fishing till there's no fish left, or until I die, which ever comes first!

Tom and Norma Peters.

A Sad Postscript:

I'm sorry to say, the latter came first for Tom. He passed away on 7/02/07 before his story could be told. He and Norma were such a help to me compiling information for this book to commemorate Suamico's fishing past. Now sadly, he's a part of it, but way before his time. He was only 61.

Ironically, he was doing what he'd set out to do; "fish until there's no fish left or die trying." He was aboard his beloved boat, the Bucky D., doing exactly that when his time came. He was doing what he enjoyed most in life, commercial fishing.

His new partner, Mark Maricque and his son Nick, along with Paul Brunette went looking for Tom when he didn't return home as scheduled. They found him north of the entrance light slumped at the wheel of his boat. Mark said "he'd have done the same for me." They shared more than a partnership.

The Tuttles and the Rodericks

Charles Tuttle and his three sons, Ray, George and Guy, all were involved in Suamico's fishing past in different ways.

Ray Tuttle, and his son Donald began fishing together before WWII started and again after Donald returned from serving his country. At some point in the 50s they received a permit to fish for sheephead* in Lake Winnebago at the request of the DNR. The sheephead had over populated the Lake and their solution was to allow commercial fisherman in to see if they could help alleviate the problem. Sheephead is a rough fish and were sold to local mink ranches for mink feed. They fished there until Ray retired and Donald was awarded the job of Postmaster for Suamico. Donald continued to fish part time with his cousin George Tuttle Jr. and his brother in law Billy Roderick.

Billy Roderick and George Tuttle Jr. aboard the "Jerry."

Donald Tuttle.

George Tuttle Sr. was in business with his father Charles for a time. They sold the business to Butch Drzewiecki and George Sr. started Northern Fisheries. The building later became Suamico's town hall and still remains next to the fire station.

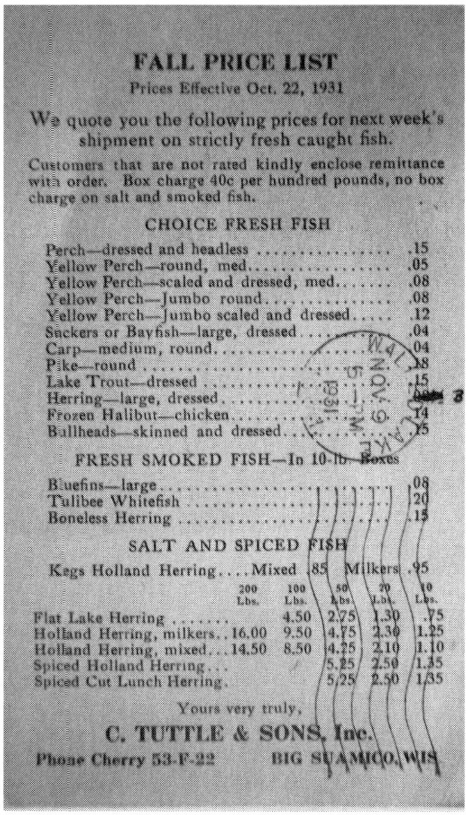

FALL PRICE LIST

Prices Effective Oct. 22, 1931

We quote you the following prices for next week's shipment on strictly fresh caught fish.

Customers that are not rated kindly enclose remittance with order. Box charge 40c per hundred pounds, no box charge on salt and smoked fish.

CHOICE FRESH FISH

Perch—dressed and headless	.15
Yellow Perch—round, med.	.05
Yellow Perch—scaled and dressed, med.	.08
Yellow Perch—Jumbo round	.08
Yellow Perch—Jumbo scaled and dressed	.12
Suckers or Bayfish—large, dressed	.04
Carp—medium, round	.04
Pike—round	.08
Lake Trout—dressed	.15
Herring—large, dressed	[illegible] 3
Frozen Halibut—chicken	.14
Bullheads—skinned and dressed	.15

FRESH SMOKED FISH—In 10-lb. Boxes

Bluefins—large	.08
Tulibee Whitefish	.20
Boneless Herring	.15

SALT AND SPICED FISH

Kegs Holland Herring....Mixed .85 Milkers .95

	200 Lbs.	100 Lbs.	50 Lbs.	20 Lbs.	10 Lbs.
Flat Lake Herring		4.50	2.75	1.30	.75
Holland Herring, milkers	16.00	9.50	4.75	2.30	1.25
Holland Herring, mixed	14.50	8.50	4.25	2.10	1.10
Spiced Holland Herring			5.25	2.50	1.35
Spiced Cut Lunch Herring			5.25	2.50	1.35

Yours very truly,

C. TUTTLE & SONS, Inc.

Phone Cherry 53-F-22 BIG SUAMICO, WIS.

Guy Tuttle's contribution was the propeller repairman. Sooner or later you were going to hit something with your prop and Guy could make it well again. He'd braze it up, straighten it out, and balance it. When he was done, it was as good as new.

Guy's son Jimmy, fished with Ray Tuttle and Ray's son Carroll while Ray's older son Donald was in the service during WWII. When Donald returned, Jim went to work for Butch Drzewiecki at the Suamico Fish Co. and then Northwest Engineering.

Ed Roderick was Art Devroy's long time fishing partner until an untimely automobile accident took his life. His daughter Mabel married Ray Tuttle's son Donald. Mabel was Brown County Clerk of Courts for many years. Ed's son Bill retired from the Post Office and then fished with Donald and George Tuttle Jr.

I went out with Donald and Ray on Lake Winnebago once. Besides the sheephead there'd be large sturgeon and catfish in the net that needed to be thrown back. The sturgeon had to be tagged first so the DNR could keep tabs on them. I was attempting to throw about a 25 lb. catfish out when it slipped out of hands and landed on my toe. I was wearing leather street shoes and overshoe boots. The prong on the side of its head went through the boot and shoe and into my big toe and broke off. It was in at such an angle that I couldn't back my foot out. I had to cut both the shoe and boot off to get at the prong still in my toe. Needless to say I wasn't very happy but Donald and Ray had a pretty good laugh.

*Sheepshead is the proper pronounciation for the freshwater drum we refer to locally as sheephead. It's also a pet name for our favorite card game, schofskopf.

Ed and Barb White

In his early years Ed got introduced to commercial fishing working with Walter and Junior Brunette. After driving truck for many years, at times hauling fish for Mike Buckarma, he bought his own gill net boat, the "Mary Jean," and he and his wife Barb (Kozloski) went fishing on their own. Soon after, Barb stayed home to establish a fish house. Ed's brother Don and later Arnie Lewis and Bob Dickinson worked the boat with Ed.

The fish house had humble beginnings. It started out as a 12 x 14 garage with a dirt floor, a canvas for a door and a torpedo heater supplying the heat. Extension cords supplied the electricity and the fish were scaled by hand.

A year later the garage had a cement floor, a furnace, a cooler and walk-in freezer complete with an ice cube machine. And an automatic fish scaler. In 1975 it had grown to become "White's Fisheries."

Their first customer was Jim Flanagan's Resort in Door County. The word spread with the quality of their product. They were soon buying fish from 17 fishermen, had six full time filleters and three part time packers.

Besides delivering fish to their own customers, White's Fisheries also processed fish for Seaway Foods, Anchor Fish, and Maricque's.

Barb became the full time bookkeeper and did whatever was needed to keep the operation going. It was non-stop from May until the Bay froze over.

The Whites retired in 1999.

Barb White.

The "Mary Jean."

The "Mary Jean" was recently sold and is being turned into a Tiki bar in Port Washington, Wisconsin. It's kind of a sad ending for such a proud vessel. Like the racehorse being relegated to pulling the milk wagon. So if you're ever in the vicinity, look her up and raise a glass to the old girl, like I did when I found her.

Getting stuck on a bar is one thing; being turned "*into*" one is another!

Norm Williquette and John Valentine

My father, Norm Williquette, and my uncle John Valentine began fishing together in the early 1940s. Uncle John had the boat and nets and needed a partner. He was married to my Dad's sister, Mercedes, so it was only natural they'd get together. They fished until the late 40s when the partnership dissolved. The perch had dropped off and relying on fishing full time to support a family became increasingly more difficult. I can recall my mother saying they had a $17 grocery bill at Martells, the local ma and pa grocery store, and didn't know how they were going to pay it. Uncle John went to work for the local fish houses and eventually became a carpenter. My Dad fished for a while with Francis Euclide and Harold (Dulie) Devroy before getting a regular job in Green Bay.

The "Betty Lou."

John Valentine's boat was the "Betty Lou". They docked at George Ladrows about an eighth mile east of the County J Bridge. In the picture, sitting on the cabin, is Lyle Valentine Jr., steering is John Valentine and standing in the rear is Norm Williquette. I can remember when the picture was taken. I was only 5 or 6 years old but I was so in awe of Lyle Jr. sitting on that cabin. I couldn't wait till I was old enough to do that. Also in the picture, across the river, is Francis Euclide's twine yard and tar kettle. It was located on the SE corner of the intersection of County Highways J and B. John and Dads twine yard was on the NW corner.

In later years it wasn't uncommon to see uncle John and Francis Euclide sitting out in front of Francis' garage on a warm sunny day. I'd always stop and chat with them. They'd tell me stories about my Dad and the good old days when fishing was good.

Around 1950, my Dad bought a rig and partnered with Art and Bill Devroy to fish it for him. They hired Rollie Belschner to work with them and I helped out in the summer months. The boat he purchased was named The "Isle of Capri." It wasn't a Lince or Lade boat and wasn't very pretty, but it served us well. That lasted until his untimely death at age 45 in 1956. Shortly there after, my mother sold everything. She needed a way to support our family so she went to Beauty School and became a licensed beautician and operated "Madeline's Beauty Shop" in Suamico for many years.

My brother Mike fished with Dick Devroy, Bill Dickinson and Reggie Netoles, and Johnnie Bender before joining the Army at age 17. My brother Jim (Big) preferred music and was part of a very successful regional band called "Short Stuff."

The "Isle Of Capri."
Art Devroy sitting on stern with his son-in-law Al Gianetta.

The Wright Bros and Freddy Valentine.

Freddy Valentine went a different route to his involvement in commercial fishing. He took his fishing roots and left Suamico for work in the big city. After a brief stint in Chicago he ended up in Milwaukee. There for a sideline he opened up a seafood store. He was doing well but the young families heartstrings drew them back to Suamico. He always claimed he could have been a rich fishmonger if sentimentality hadn't prevailed. He went to work for Northwest Engineering.

Shortly before he retired he bought Otto Marks' fishing rig. He rebuilt the "Eugene M." and fished in the evenings with his son Tom and Tom Peters until he retired from Northwest. He never did get to fish fulltime, illness cut short his dreams.

The "Eugene M." parked next to their house on the Suamico River. The kids belong to Ronnie and Loretta Devroy.

His wife Leila was also born into a fishing family. Her father Martin Wright and his brother Harvey were long time Suamico fishermen. Her brother Clifford and his son Dale took over the brothers operation when they retired. Dale fished into the 90s.

Freddy and Leila were known for their great sense of humor. Freddy's favorite saying was "I couldn't do any better and she couldn't do any worse."

Freddy and Leila had three sons; the oldest, Glenn, went to Alaska to work for the railroad, Larry became an Episcopal Priest and Tom was a truck driver.

Larry and I were constant companions when we were kids. Every Christmas we'd get a new gun and holster set and high-top leather boots with a pocket on the side for a jack knife. That's all I can ever remember asking for. Larry had a collection of every "Tarzan" comic book dating back to 1948.

Larry's aunt, his dad's sister Mabel Campbell, gave us each an old Civil war rifle. When I visited him recently at his home in Kansas, he still had his. I must have traded the one I had to him for something, because now his brother Tom has it.

She also gave us a beautiful old wind up Victrola that we brought out to the "big boys shack." It was a shack out in the woods behind Art Fiest's that was built by the older boys, Glenn Valentine, Dale Leanna, Bobby Dickinson, Ora Cahoon and others. When they outgrew it, Larry and I inherited it and then my brother Mike and Larry's brother Tom and so on. The Victrola didn't last a week before somebody stole it.

Freddy Valentine getting the "Eugene M." ready to do some serious fishing.

The Zakowski Brothers

The Zakowski Brothers, Arnie and Herb, started fishing together when Herb returned from the service after WWII. Prior to that, Arnie fished with George Lince.

Estelle posing on the "Tateter Z."

Their first boat was named the "Tateter Z." after their sister Estelle. It was her nickname. Around 1950 they commissioned George Lince to build the "Judy," named after their sister Julia who was always known as Judy. If another boat had been built it would have been named the "Marie" after Arnies wife. (The "Judy" is in the painting on the cover.)

The "Dianne." Herb is leaning out of the window.

Besides fishing drop nets they also fished gill nets. Their gill net boat was the "Dianne." Dick Devroy tells the story of going out on the maiden run of the "Dianne" with a bunch of other fishermen and "celebrating" way too much. He was piloting on the way in and ran aground coming in the river.

To make matters worse, he was expecting steak when he got home but didn't make it on time. I happen to be playing cards with his wife Teresa that day and when he didn't show up for suppertime, she cooked up his steak and told me to eat it. He came home late and was stumbling around looking for his steak. He woke Teresa up. She hollered from the bedroom "Danny ate it." To this day he hasn't forgiven me.

Once Arnie and Herb were fishing out of Peshtigo through the ice and their snowmobile fell through while trying to cross the crack. They got someone to dive down and tie buoys to it so when the ice went out in the spring they'd be able to locate it. When the ice went out, they found it, tied ropes to it and pulled it as close to shore as they could with their fishing boats, and the rest of the way with their truck. After a little TLC, it started right up.

Zakowski's old fishing truck.

SOUTHERN GREEN BAY COMMERCIAL FISHERMEN'S
ASSOCIATION'S BANQUET - 1989

Deposited proceeds from banquet (Orchestra paid in cash)		$5108.00
Expenses for banquet		
Val Drzewiecki	330.00	
Maricque's Bar	100.00	
Rock Garden	2771.70	
Seaway Foods	35.50	
Fiala Fish Market	16.50	
Boats & Us	787.50	
Seaway Print ing	11.03	
Schroeder's Flowerland	143.17	
	$4195.40	

Transactions for part of year 1989			912.60
Balance on hand as of 1/1/89			3734.77
Paid Wisconsin Commercial Fishermen's Assoc.	1717.98		
Paid Jim Maricque (expenses for trip to Madison)	87.28		
Deposited dues received for 1989		775.00	
Balance as of 6/23/88 in checking acct.			3617.11
Balance as of 6/23/88 in savings acct.			582.18

Minutes of meeting March 23, 1989

Meeting called to order by President James Maricque.
The first item of business: Jim Maricque told the DNR personnel that our group was not in favor of a minimum harvest to keep a license. No hearing has been called by a Legislative Committee so far. Kincaid is the chairman of the Senate Committee. The Assembly Committee is chairedby Black from Milwaukee. Two day meeting in Madison last week on the Rules Package. Goyke feels that 3 items could be amended. Meeting will be in Green Bay Tuesday at 4:00 P.M. in Green Bay at Maricque's Bar. This meeting will be the Wisconsin Commercial Fishermen's Assoc.

Quota increase was submitted last week to go to the DNR Board at the May meeting. This will have to go the whole 9 month routine. Jim Maricque says that the quota increase will be a fight with the sportsmen. The Green Bay Sportsmen Group will oppose this.

There is also discussion to move the Southern Green Bay line further north. This is only discussion now.

Also read a letter from Elaine Johnson concerning Buyout Legislation.

Dennis Keller made a motion that we decide if we are in favor or opposed to a compensation package proposed by Elaine Johnson. Seconded by Tom Peters and Don Tuttle. FOR: 1 AGAINST - 23. Jim Maricque - abstained.

Glory Hermes discussed banquet plans. We have a boat to sell and hope attendance will be good.

ELECTION OF OFFICERS: Van Drzewiecki moved to keep same officers. Seconded by Don Tuttle.

Motion to adjourn by Don Tuttle. Seconded by Russell Devroy. Meeting to be held in May, July, September and November.

The Quota System on the Bay of Green Bay

1982 May	No Limit
1983 Feb.	200,000 lbs.
1984 Feb.	350,000 lbs.
1985	350,000 lbs.
1986 July	400,000 lbs.
1987	400,000 lbs.
1988	400,000 lbs.
1989 Feb.	400,000 lbs.
1990 Sept.	475,000 lbs.
1991 Feb.	400,000 lbs.
1992	400,000 lbs.
1993	400,000 lbs.
1994 June	300,000 lbs.
1995 Apr.	300,000 lbs.
1996 Sept.	300,000 lbs.
1997 Mar.	200,000 lbs.
1998	200,000 lbs.
1999 Feb.	200,000 lbs.
2000	200,000 lbs.
2001 Oct.	20,000 lbs.
2002	20,000 lbs.
2003	20,000 lbs.
2004 Feb.	20,000 lbs.
2005 Dec.	60,000 lbs.
2006	60,000 lbs.
2007	60,000 lbs.

THE LINEN THREAD CO., INC.
MANUFACTURERS OF FINE NETTINGS
FEBRUARY 1957
SUN MON TUE WED THU FRI SAT
1 2
3 4 5 6 7 8 9
10 11 12 13 14 15 16
17 18 19 20 21 22 23
24 25 26 27 28
JANUARY 1957
MARCH 1957